Brick by Brick

Brick by Brick
Participatory technology development in brickmaking

Kelvin Mason

Practical Action Publishing Ltd
25 Albert Street, Rugby, CV21 2SD, Warwickshire, UK
www.practicalactionpublishing.com

© Intermediate Technology Publications 2001

First published 2001\Digitised 2013

ISBN 10: 1 85339 529 3
ISBN 13 Paperback: 9781853395291
ISBN Library Ebook: 9781780442655
Book DOI: https://doi.org/10.3362/9781780442655

All rights reserved. No part of this publication may be reprinted or reproduced or utilized in any form or by any electronic, mechanical, or other means, now known or hereafter invented, including photocopying and recording, or in any information storage or retrieval system, without the written permission of the publishers.

A catalogue record for this book is available from the British Library.

The authors, contributors and/or editors have asserted their rights under the Copyright Designs and Patents Act 1988 to be identified as authors of their respective contributions.

Since 1974, Practical Action Publishing has published and disseminated books and information in support of international development work throughout the world. Practical Action Publishing is a trading name of Practical Action Publishing Ltd (Company Reg. No. 01159018), the wholly owned publishing company of Practical Action. Practical Action Publishing trades only in support of its parent charity objectives and any profits are covenanted back to Practical Action (Charity Reg. No. 247257, Group VAT Registration No. 880 9924 76).

Reasonable efforts have been made to publish reliable data and information, but the author and publisher cannot assume responsibility for the validity of all materials or for the consequences of their use.

The manufacturer's authorised representative in the EU for product safety is Lightning Source France, 1 Av. Johannes Gutenberg, 78310 Maurepas, France. compliance@lightningsource.fr

Contents

	Acknowledgements	vii
1	INTRODUCTION	1
	The shelter crisis	1
	The choice of brick production	3
	ITDG and the Shelter Programme	7
	About this book	10

PART I APPROPRIATE TECHNOLOGY TRANSFER

2	BRICKMAKING – THE TECHNOLOGY AND THE PRODUCT	13
	What *exactly* is a fired-clay brick?	13
	A well-known technology?	14
3	APPROPRIATE TECHNOLOGY AND ITS TRANSFER	22
	Appropriate technology – in search of a working definition	22
	Defining technology transfer	25
	Elements of technology transfer	28
	Case studies in technology transfer	31
4	PARTICIPATORY TECHNOLOGY DEVELOPMENT	34
	Enhancing knowledge, skills and choices	34
	Ownership through participation	34
	PTD as a project tool	35
	Conclusion	46

PART II THE BRICKMAKING TECHNOLOGY TRANSFER PROJECT

5	THE CHOICE OF PERU AND ECUADOR	49
	Why Peru?	49
	The project objective	51
	The situation in Peru	52
	The situation in Ecuador	54
6	EARLY EXPERIENCE IN ZIMBABWE – THE COAL-FIRED CLAMP	56
	Brickmaking in Zimbabwe – the historical context	56
	The choice of the coal-fired clamp	58
	The process of technology transfer	59

	Successes and feedback	60
	Lessons learned	61
	Conclusions – but not the end of the story	64
7	THE START OF THE PROJECT IN PERU – GATHERING DATA ON ENERGY EFFICIENCY	65
	Designing the technology transfer project in Peru	65
	The draft technology transfer plan	70
	The importance of measuring energy efficiency	71
	Methodology for measuring the energy used to fire clay bricks	73
8	THE PROJECT IN PERU AND ECUADOR – EXPERIMENTATION AND CONSOLIDATION	83
	An inauspicious start – the Cajamarca coal-fired clamp	83
	Gathering together Latin American experience	84
	The project in La Huaca	85
	More fuel alternatives: coal briquettes and waste engine oil	89
	Developing hardware – crushers, extruders and oil-burning engines	93
	Training through exchange visits	94
	The Ecuadorian perspective	96
	Institution building – brickmakers' associations	98
	Ten rules for energy-efficient, cost-effective brickmaking	99
9	PROJECT OUTCOMES	105
	Training: getting it right	105
	Sustainability: what next?	106
	Monitoring technological change	106
	Environment: more to be done	109
	Information dissemination: getting the message across	110
10	GUIDELINES FOR PARTICIPATORY PROJECTS	111
	Formulation: the project proposal and work plan	111
	Summary: consolidating guidelines	113
	Further reading	116
	References	120
	Appendix 1 Project time frame	123
	Appendix 2 Energy monitoring forms	127
	Appendix 3 A comparison of calorific values for different fuels	131

Acknowledgements

Many people have worked on the technology transfer project in Peru, Ecuador, Britain and Zimbabwe. Their hard work is appreciated and thanks are due to them all. The following list includes only those who have directly helped the author in the preparation of this book: special thanks to them for their contributions, insights and feedback.

Juan Francisco Coronado Acaro (Brickmaker, La Huaca, Peru)
Ray Austin (Consultant, UK)
Alfredo Barriga (ESPOL, Ecuador)
Valentin Coronado Espinoza (Brickmaker, La Huaca, Peru)
Victor Carmen Garcia (Brickmaker, La Huaca, Peru)
Mauricio Gnecco (ITDG Peru)
Martha Jara Huayta (Brickmaker, Ayacucho, Peru)
Mario Jara (Consultant, Peru)
Lucky Lowe (Project Manager, ITDG UK)
Steve Murambidzi (Brickmaker, Harare, Zimbabwe)
Emilio Mayorga Navarro (Project Manager, ITDG Peru)
Guillermo Pincay Romero (Consultant, Ecuador)
Luis Alvarez Rodriguez (Oil-burner, Trujillo, Peru)
Jorge Marquina Ruiz (ITDG Peru)
Otto Ruskulis (ITDG UK)
Theo Schilderman (Senior Sector Specialist, ITDG UK)
Clare Tawney (Editing)
Peter Tawodzera (ITDG Zimbabwe)

1
INTRODUCTION

The shelter crisis

EVERYONE NEEDS A HOME. People need roofs over their heads for protection from the elements. They need security of tenure; they need a safe and healthy environment in order to live full lives. A home is somewhere to raise a family; in many cases it is also a place to work. Thus, shelter is not only about satisfying a basic need, it is intimately connected with livelihoods and economic development. For many people, however, satisfying this requirement for shelter is a very difficult proposition. The problems faced by home seekers are wide ranging. Often land to build on, housing finance and affordable building materials are almost impossible to obtain.

The world is facing a shelter crisis; and it is at its most acute in the countries of the developing world. The statistics make for gloomy reading. At a conservative estimate, over one-fifth of the world's population – more than one billion people – are homeless or inadequately housed. Predictions for the future are even gloomier: according to the United Nations Centre for Human Settlements (UNCHS, 1996), this number is set to double by 2010.

Population growth and increasing urbanization exacerbate the problem. In the face of declining rural economies, people continue to flood into towns in search of jobs. Large informal settlements, slums or squatter camps surround almost every major population centre in the developing world. In many cases, the area and population of these settlements exceed those of the formal town or city itself. *Newsweek*'s first edition of the millennium forecasts that from 1995 to 2025:

- the percentage of the world living in urban areas will rise from 45 to 61 per cent;
- the number of slum dwellers will increase from 810 to 2100 million;
- but, conversely, the number of people with *no* access to shelter will fall from 1500 to 700 million (seeming to imply that slums will come to be considered adequate shelter).

(*Newsweek*, 2000, quoting from *Encyclopaedia of the Future*)

Meanwhile, the population in rural areas, too, will continue to grow. This means an increasing need for land, shelter and building materials.

However the problem is viewed, all statistics indicate that very little impact is being made on the world's huge housing deficit. As late as the 1980s, governments in some developing countries conceived rallying calls such as 'Shelter for all by the year 2000!' In retrospect, the exclamation mark appears sadly ironic. In relation to the escalating problem and the ineffective global response, the contribution of non-government organizations (NGOs) has been very small. Nevertheless, there are many examples of good practice that can inspire those willing to learn. There have been successes, and the housing situation in numerous communities has been improved. The challenge now is to disseminate these experiences and promote the adoption of shelter solutions on a mass scale. Building houses, assisting people to build, attempting to increase building materials production and choice, and lobbying for more appropriate building standards: all these activities can help to pave the way forward.

In view of the daunting size of the crisis, however, it is tempting just to give up. When the solution involves building low-cost housing over vast tracts of countryside, and increasing the sprawl of towns and cities, we may wish to turn our back on the problem. This option is not open, however. From whatever perspective the problem is viewed, be it humanitarian, communitarian or through the pragmatic lens of self-interest, the shelter crisis cannot be ignored. The slum areas that surround most Third World cities are a cause of suffering not only to their inhabitants: they are also a threat to the nearby enclaves of the better off; and they are a blight upon the economic development of the whole city. Bringing people into secure accommodation in safe and sanitary neighbourhoods (the so-called three S's, which are the aim of many shelter programmes) also opens up new markets among them for products and services from both local and international companies. Business cannot keep catering only for an élite, which is declining percentage-wise, who can afford their products and services.

Just as the environment cannot be preserved with a quick fix, so the shelter crisis requires a long-term view and *will* ultimately have to involve rich people in developed countries making sacrifices in their standard of living. The fact is that the solution cannot be found in the developing world alone. The world as a whole cannot afford the energy-intensive 'first world' lifestyle that it is being encouraged to seek: we simply do not have the resources – or the sinks for the quantity of waste that would be generated. So, there will have to be major changes in thinking. The shelter situation might be compared to food shortages in the world: there may be enough food to go around; but the problem is one of distribution and excess. We must not be afraid of saying that the shelter crisis is intolerable and that the individual must play his or her part to change things. We have a moral obligation to be courageous on behalf of the homeless and the ill-housed. And their overwhelming numbers make it imperative to develop new and creative shelter solutions.

The trend towards a more open global economy has meant that governments in the developing world have been obliged to be more market oriented and export driven. Public sector budgets have shrunk, and this has meant that state support for the social sector has declined. An example of this is in Peru, where the Housing Ministry and Housing Bank were actually closed in the early 1990s. Increasingly, the provision of shelter is being left to NGOs and the private sector; and their efforts are not enough to make more than the smallest impact on the problem.

Given these arguments, it is evident that the problem certainly cannot be solved by technology alone – appropriate or otherwise. Technology is only one component of the solution. In most cases the technology – or at least the knowledge component of technology – exists. Given the opportunity, we can choose, produce and use a variety of appropriate roofing, walling and binding materials. Naturally, there are always improvements possible in costs, energy efficiency and environmental preservation. Generally though, we have enough technology. The problem is getting it to where it is needed on the scale that is required. What is needed is to be able to apply what we know in an environment where it can have sufficient effect.

The brickmaking project discussed in this book is only one of many initiatives that technology transfer organizations like ITDG are taking. The foundation of the project is participation: communities defining their own needs and solutions, with NGOs acting as their staunchest allies. Education is the principal element of the work. Increasing knowledge is the key to empowerment, self-determination and sustainability. And that is true for both hemispheres. Education is vital in the First World too, so that developmental issues may be understood and people may be motivated to act.

The choice of brick production

What is the contribution that improved brick production makes to the shelter crisis? The short answer is that houses cannot be built without materials, and fired-clay bricks are a durable and popular choice. To work in the shelter sector without promoting the increased supply of appropriate, affordable materials would not, ultimately, be viable. The production of building materials creates jobs and generates income: it can be an engine for economic development. There is an argument against promoting the production of fired-clay bricks and other 'modern' building materials, however. As we shall see, these materials are energy-intensive; and the energy most often comes from non-renewable fuels. One possible strategy towards more sustainable construction is to promote the use of alternative, low-energy or renewable materials, such as earth or bamboo. The production of simple earth blocks only requires around a thousandth of the energy needed to fire an equivalent quantity of bricks. Even when the blocks are stabilized with cement, the proportion of cement used is never more than a sixth of the total weight of the building material produced. Wherever possible and appropriate, these alternatives should be considered for projects.

There are limits to the usage of these materials, however. A combination of factors, including mass urbanization, consumer attitudes and inappropriate building standards are generating an increasing demand for conventional materials such as steel, bricks and cement, which unfortunately do have a substantial negative environmental impact. This trend is hard to reverse and it is therefore necessary to adopt a second strategy, aimed at mitigating this environmental impact. What this impact may be is discussed in the following section, but at this point we turn from the technology itself to the people most affected by it: brickmakers.

Reasons for working with small-scale brickmakers

Informal sector and small-scale brick production is important for poor people in many developing countries. In India, for example, around 2 million people are directly employed in brickmaking; in Bangladesh it is estimated that over 200 000 people are employed in the sector. Demand for bricks continues to rise steadily in most developing countries. Satisfying this demand can create much-needed jobs in economies where unemployment, underemployment and poverty are huge problems. Furthermore, small-scale brickmaking can create these jobs in peri-urban, urban fringe and rural areas where the need is often greatest. Creating employment outside urban areas can also decrease the continuing rural–urban drift, whereby people stream into cities looking for work – work that is often not available.

The environmental impact of brickmaking

There is also a case for working with brickmakers that stems from energy and environmental concerns. Brickmaking is an activity that can have considerable negative impact on the environment. In the first instance, agricultural land may be degraded due to clay extraction, leaving behind it a scarred and unproductive landscape. Meanwhile, indigenous biomass resources or fossil fuel reserves are depleted when bricks are burned. In most instances in the developing world, it is indigenous biomass resources that are under the most pressure. The most common fuel used to burn bricks is wood – or sometimes its derivative, charcoal. There are very few instances where the fuelwood used is being extracted from managed woodlots or where trees are being planted to replace those felled to burn bricks.

All burning processes generate greenhouse gases, which – in excessive quantities – are detrimental to the earth's atmosphere, and brickmaking is no exception (see Box 1.1). In India for example, brickmaking accounts for around 6 per cent of the country's total carbon dioxide emissions. The issues of fossil fuel depletion and greenhouse gas emissions are becoming increasingly important on the global agenda. The Kyoto Protocol on global action to reduce greenhouse gases comes into effect from 2008 to 2012. Policy makers are gradually beginning to realize that it is not only large-scale, formal-sector industries that need to take action to reduce emissions. The combined emissions from small-scale producers and the informal sector are also globally significant. Meanwhile, smaller-scale producers operate under greater constraints than large-scale industries: their profit margins are slimmer and their ability to invest in change much less. Developing effective policies and processes that can work for the small-scale sector is therefore likely to be an even more challenging task.

Where brickmaking is concerned, small-scale producers are relatively more damaging in terms of greenhouse gas emissions than their large-scale counterparts. This is partly because the brick clamps used by most small-scale producers are temporary structures with insufficient insulation, and therefore very wasteful of energy, and partly because most small-scale producers burn fuelwood, which may result in deforestation if it is not taken from a sustainable source. One of the many negative consequences of deforestation is the reduction in a carbon sink (see Box 1.1).

This may suggest that small-scale brickmaking should be discouraged in favour of large-scale production. Pragmatism directs action once again here. Just as alternative building materials such as bamboo can only provide a very small part of the solution, so supplies by large-scale producers are limited in their coverage, and small-scale producers are the only viable sources of bricks in many rural parts of the developing world (see Box 3.2 for a discussion of the difficulties facing large-scale brickmakers in Ghana). So instead of discouraging small-scale brickmaking, projects such as the one described in this book try to promote more fuel-efficient technologies by brickmakers.

This is an objective largely shared by small- and medium-scale brickmakers, who are themselves at risk from the depletion and increasing cost of fuelwood. Legislation to prevent deforestation and protect the environment is in effect or coming into effect in many countries. Brickmakers suffer from the enforcement of such legislation, being denied access to the fuel they need to burn bricks. Ultimately, many such brickmaking operations may prove unsustainable. Those brickmakers who are able to introduce, develop and adapt improved, more environmentally friendly production techniques, however, can greatly increase their chances of commercial survival. By increasing their fuel efficiency and using alternatives to wood, they can maintain or even expand their operations, supplying the increasing demand of the construction sector for bricks.

Some industrial and agricultural processes produce large quantities of combustible wastes or residues. Some of these could potentially be used as fuels in brickmaking, as a substitute or partial substitute for fuelwood or fossil fuel. In the main, these residues are not used for productive processes; they are disposed of either by dumping or burning. Research into the

Box 1.1 Greenhouse gases, the greenhouse effect and climatic change

It is sometimes forgotten that life on earth exists because of the 'greenhouse effect'. It warms the planet's surface by 33°C and raises the temperature of the background atmosphere to 15°C at the surface. This natural warming is caused by the build-up of so-called 'greenhouse gases', which trap heat from the sun inside the earth's atmosphere. It has allowed water – the basis of biological evolution – to exist on the earth's surface. Without the greenhouse effect, the temperature of the earth's surface would be –18°C.

The main greenhouse gases (GHGs) are carbon dioxide (CO_2), methane (CH_4), nitrous oxide (N_2O) and chlorofluorocarbons (CFCs). Increased emissions of GHGs due to human activity threaten to amplify the natural greenhouse effect. The additional GHGs emitted may heat the planet by up to 5°C in the next 50–100 years, resulting in so-called global warming. Human-induced emissions may increase climate change to a rate that is too fast to allow some species to adapt and will put pressure on human societies. Hence, the life-giving greenhouse effect becomes the life-threatening greenhouse problem.

Global warming will not have the same effect all around the world. It could, however, dramatically alter regional climate regimes. Warming will be greatest at the poles, up to three times the global average, and least in the tropics. One effect of global warming will be a rise in sea levels. If current emission trends continue, sea levels could rise by as much as 40 cm by the year 2050.

The theory of the greenhouse effect has never been proven. Nevertheless, empirical evidence to support the theory is readily available and robust. Although the earth has survived temperature changes of similar magnitude in the past, these have generally occurred over 5–15 millennia, rather than over 5–10 decades. It is the rate of change as much as the magnitude of the change that will determine the extent of damage to ecosystems and human societies.

Human-induced emission of carbon dioxide, the most problematic GHG in terms of quantity, and hence effect, were equivalent to approximately 7 billion tonnes of carbon in 1990. This represents a global average emission rate of 1 tonne of carbon per person. One principal source of carbon dioxide emissions is the combustion of fossil fuels – oil and coal – and the manufacture of cement, amounting to 5.6 billion tonnes of carbon per year. Deforestation and other forms of land use change result in up to 2.5 billion tonnes of carbon per year. Not surprisingly, the developed countries have by far the highest per capita carbon dioxide emissions.

Most spheres of modern human activity cause the increasing rate of GHG emissions. The most culpable are, perhaps, energy supply, industry, buildings, transport, agriculture, waste disposal and forestry. CFCs are only produced by industrial processes. The principal natural sinks for carbon dioxide are photosynthesis by trees and other green plants, absorption in the ocean, uptake by soils and – unfortunately – build-up in the atmosphere. There are virtually no biological sinks for methane, nitrous oxide or CFCs.

Source: Economics of Climate Change: Implications for National Development, USAID, 2000

use of such residues in brickmaking and other building materials production is an important area of interest.

With relatively minor improvements in brickmaking practices, many brickmakers could increase their fuel efficiency. They could improve their production processes, particularly the drying and burning stages, to either save fuel or produce more thoroughly baked bricks, ensuring a better-quality product that could command a higher price. Poor product quality is often a critical issue for small-scale and informal brickmakers: in general, it means that their bricks are unsuitable for urban or industrial construction and hence they are reduced to selling at a low price.

For small-scale brickmakers in the developing world, the cost of fuel is typically half the cost of production. In parts of Asia, for example, it is estimated that up to 50 per cent of the production costs of small-scale brickmakers is the cost of fuel (Koopmans and Best, 1993). Meanwhile, in Sudan, the cost of fuel can be up to 60 per cent of total production costs (Bairiak, 1999). These figures are typical of the sector. It is apparent, then, that any improvements in fuel usage could have a positive impact on the income of brickmakers or serve to reduce the cost of their bricks.

Opportunities for innovation among brickmakers

Brickmakers are acutely aware of the problem of fuelwood scarcity and increasing price. Generally, they are also aware of the threat that deforestation poses to the local and global environment. Some do try to address these problems: they do experiment and innovate by themselves. And sometimes these innovations result in significant improvements in production, notably fuel efficiency. For example, the brickmakers of La Huaca in Peru, who were suffering from particularly stringent restrictions in fuelwood supply, had already begun to experiment with rice husk as a fuel before they were approached by the ITDG project (see Chapter 8). Many small-scale and informal sector brickmakers, though, are not in a position to instigate systematic production trials, modifying a range of variables in a controlled way. In addition, they might justifiably shy away from making drastic changes to traditional practices just because an expert or a book recommends such action. The fear would be that a whole batch – perhaps many thousands of bricks – could be lost. Very few small-scale brickmakers could afford to take such risks or stand such a loss.

Information on innovations in small- and medium-scale brickmaking is scarce. Furthermore, very little of that information actually reaches the brickmakers themselves, and still less in a form and format that would be directly useful to them. Brickmakers can and do, however, learn from each other, especially where a number are gathered together in a particular area or where a producer's association or other form of collective organization is active. The opportunities for small-scale brickmakers from different districts or countries to learn from each other are, however, severely limited.

Comparatively few NGOs, government departments, research groups or enterprise support organizations are working directly with brickmakers in the developing world (the work of the few that are is summarized in Box 1.2). This is perhaps surprising given the number of people who rely on the sector for their livelihoods and the environmental impact of brick production. Such organizations could play an important role in facilitating information exchange between brickmakers, supporting research and innovation, promoting institutional development, skills training, and encouraging dialogue between brickmakers and policy makers.

> **Box 1.2** International organizations involved with small-scale brickmaking
>
> Building materials and construction technologies that are appropriate for developing countries are being developed and documented in many parts of the world. To disseminate information on these technologies, a number of organizations coordinate their work through the Building Advisory Service and Information Network (BASIN). They include: the Shelter Forum, Kenya; GATE–GTZ, Germany; ITDG, UK; SKAT, Switzerland; and CRATerre–EAG, France.
>
> The BASIN member specializing in walling materials is GATE–GTZ. They are a leading agent in the world of small-scale brickmaking; they have commissioned action research and publication of many technical briefs on the subject of brick production and firing technologies, alternative fuels, equipment providers and so on.
>
> The Swiss development agency SKAT, funded by SDC (Swiss Agency for Development Co-operation), has been involved with transferring the vertical shaft brick kiln (VSBK) from China to India (this is referred to at the end of Box 4.1). They have collaborated with several institutions in India in measuring the energy efficiency of this kiln, and in experimenting with various transfer methodologies to better understand the kind of institutional set-ups that facilitate the capacity-building process.
>
> The Stockholm Environment Institute (SEI) has also undertaken a considerable amount of work on the small-scale production of building materials, with a particular focus on energy efficiency, including lime and brick firing. ITDG's partner in Ecuador for the brickmaking project, Alfredo Barriga, had previously worked with SEI on brickmaking projects. SEI has been funded by SIDA (Swedish International Development Authority), UNDP, UNEP, UNIDO, FAO and the World Bank.

ITDG and the Shelter Programme

Within its Shelter Programme, the non-governmental organization Intermediate Technology Development Group (ITDG) works to encourage brickmakers to innovate their technologies and exchange information on energy-saving methods. Brickmaking is one part of the Shelter Programme, and this in turn is only one programme within an organization covering the sectors of manufacturing, food production, agro-processing, energy, building materials and shelter, disaster prevention and mitigation, transport and mining.

ITDG's mission

ITDG's mission is to build the technical skills of poor people in developing countries, enabling them to improve the quality of their lives and those of future generations. Most of the world's poorest people earn their living from small-scale enterprises that they run from their homes, fields and workshops. They rely on small-scale, often traditional, technology to succeed. ITDG helps people to build on their existing knowledge and skills, and to improve their technical capacity in ways that are appropriate, affordable and sustainable.

Founded in 1966 by E.F. Schumacher, ITDG is now an international group of development organizations with charitable status. Its head office is in the UK, with regional and country offices in the three continents of Africa, Asia and South America.

The Shelter Programme

The Shelter Programme is one of ITDG's main programme areas. It is a long-standing programme, which has been active since around 1975. The initial focus of the programme, up to a decade ago, was on developing livelihoods from building materials production and promoting the increased supply of affordable, locally produced building materials. Since 1987, however, the programme has broadened its approach to better address the shelter needs of the more than one billion homeless or poorly housed people throughout the world, recognizing that they urgently need safe, secure, sanitary and sustainable places to live. This expanded agenda has meant that the programme has engaged in a wider range of activities. Although improving access to materials and technologies remains an objective, as does helping to create employment and generate income, other issues have become at least equally important, including access to land and finance, participation, equity and the environment (see Box 1.3).

Shelter processes have a significant role in the overall development agenda, helping to reduce people's poverty and vulnerability and increase their assets. In short, shelter processes can contribute towards establishing sustainable livelihoods. ITDG's Shelter Programme Strategy, for 2000–2005, recognizes the importance of the 'sustainable livelihoods approach' in understanding development. Within that approach, there are particular components upon which access to adequate shelter has a significant bearing. For many people, the dwelling itself constitutes a considerable part of their physical capital. In addition to protecting people from normal weather effects, an appropriate, robust dwelling can reduce their vulnerability to disasters. The process by which people acquire shelter, if undertaken at a community-based level, can help to increase people's social capital, especially in the context of increased interaction between community members and the empowerment of the community. In addition, for many poor people, especially women, the home is also their workplace. So, access to better shelter can lead to an increase in income as well.

These considerations have determined the overall directions that ITDG's Shelter Programme now follows:

- To explore how improvement to low-income shelter can make a long-term contribution to the broader development agenda; for example, by generating sustainable livelihoods, improving the living environment, empowering poor women and men, and reducing their vulnerability.
- To address, at a more practical level, how poor women and men can enhance their shelter assets and what other stakeholders can do to enable that.

ITDG's Shelter Programme currently operates from five of ITDG's seven country offices. Long-term shelter-based projects are being undertaken in Kenya, Peru, Sri Lanka, Sudan and Zimbabwe. Smaller-scale, shorter duration and generally more research-based activities have also been undertaken in a number of countries, including India, Uganda, Bolivia and Ecuador. ITDG's UK office is the focus of international dissemination and advocacy activities. ITDG is a member of the Building Advisory Service and Information Network (BASIN), which comprises nine organizations worldwide that have significant expertise and experience in the building and housing sector.

Box 1.3   The approach of ITDG's Shelter Programme

The purpose of ITDG's Shelter Programme is to develop and disseminate innovative shelter technologies and approaches with poor women and men in selected countries. For the programme to fulfil its purpose, action is required over five broad, shelter-related themes:

- *The environment.* The aim being to improve the environmental sustainability of low-income housing through improving the health of poor dwellers and the health and safety of producers; increasing energy efficiency and reducing the waste associated with production; and increasing the availability of environmentally friendly technology options.
- *Markets.* The focus being on enabling poor women and men to participate more actively and on better terms in the markets for building materials, construction and shelter. Achieving this aim will involve undertaking market surveys and analysis; seeking to reduce the constraining factors; helping the informal sector to increase its market share; supporting women's market development initiatives; and analysing the impact of technology choice and income generation from a shelter standpoint.
- *Partnerships and alliances.* The objective being to strengthen partnerships and alliances for pro-poor changes in shelter programmes and policies through the development and strengthening of formal and informal networks; facilitating information exchange; promoting a greater range of partnerships, especially between communities and local authorities; and disseminating case studies on good practices.
- *Reform.* The goal of reducing the policy, legal and institutional constraints faced by poor women and men seeking adequate shelter can be pursued via the identification and removal of inappropriate constraints, enhancing increased security of tenure, and the implementation of needs-based appropriate standards and regulations.
- *Advocacy.* The target being to make the case for shelter in improving the livelihoods of poor women and men by promoting the development and utilization of a greater range of appropriate indicators; advocacy strategy development; increasing the advocacy component of projects; and documenting and disseminating evidence of impact and change to appropriate audiences.

ITDG's Shelter Programme is expanding and developing increased capacities, especially in the social, economic and environmental aspects of building and shelter. Small-scale brickmaking remains an area of significant expertise within the programme with project activities in Peru, Sri Lanka, Sudan and Zimbabwe. Due to enhanced concerns over the significant environmental impact of brickmaking in many countries and the livelihood threat this poses to brickmakers, this aspect of the programme's work has increased in importance.

Increasing poor people's access to technology

It may not yet be clear why there is a need for NGOs such as ITDG to intervene to introduce improved technologies, including brickmaking technologies, in developing countries. If the technologies already exist, what is to prevent them spreading in ways that are normal in developed countries: through suppliers, advertising, word of mouth, etc? The answer to this lies in the fact that the world's poorest people are those who have little stake in the formal economy of their countries – and few safety nets to support them. The poor are marginalized people, isolated from information, excluded from the benefits of the global economy and overlooked by mainstream planners. Neither the forces of the state nor those of the market work effectively to meet their technology needs. They have limited access to expensive, imported technologies, or to other goods and services such as electricity, water, education, health care and financial services, including loans and savings.

New forces threaten to push the world's poor people further to the margins, locking them into a deeper poverty trap. 'Globalization' has created an aggressive market-based economy that stretches into almost every neighbourhood in the world. This rapid, sweeping change has outstripped the capacity of poor people's existing technology to provide them with secure livelihoods. Either it is not productive enough, or the goods produced are not of the right type to compete in this new global marketplace. Helping people to secure sustainable livelihoods, whether we are referring to the livelihood of an African dryland farmer or a Peruvian brickmaker, should encompass the following aspects.

- First, it means helping people to improve their incomes and the quality of their lives without damaging the environment that their children will inherit.
- Second, 'sustainability' means ensuring that the benefits of NGO projects continue well into the future. Short-term technology improvements may be quickly outpaced by the rapidity of global change. So 'appropriate technology' must be redefined: technology is only appropriate if its users are able to continue to adapt and innovate, whatever the future brings, and long after the NGO's involvement in a project has ended. This means that projects should aim not only to improve technology 'hardware', but also the 'software': to build the capacity of producers to develop their skills, knowledge and understanding of the technology options available to them. This will enable people to identify and develop their own solutions.
- Third, ordinary people will only benefit from technological change if they can sustain a presence in the market – locally, regionally and even internationally. Where private sector business support services such as market analysis, marketing training and so on are absent – as they are in many parts of the developing world – there is a need for non-government agencies to provide or to stimulate this type of support.

About this book

This book is about an ITDG project that worked with brickmakers and institutional allies in Latin America. It describes a technology transfer and development project, which facilitated the exchange of ideas between brickmakers and technologists from Peru, Ecuador, Zimbabwe, Britain and latterly Colombia. In addition, the project drew on ITDG's experience with brickmaking across the world, notably in Sudan and Sri Lanka.

Financed by the Knowledge and Research (KAR) fund of the Department for International Development of the British Government, the project was designed to foster more environmentally sensitive and operationally sustainable production in the small-scale and informal brickmaking sector. This book presents the challenges that have to be faced by those working with the sector. It also proposes possible solutions for effecting changes that could have a mass impact.

Part I deals with 'appropriate technology' and its transfer. First of all, in Chapter 2, the practical elements of how bricks are moulded and fired are described in brief. The structure of the basic brick clamp, used by small-scale brickmakers throughout the world, is described. This is followed in Chapter 3 by a discussion of technology choice, including what criteria might be considered to constitute an appropriate technology. As was mentioned earlier, the market forces of our global economy do not usually work to spread such appropriate technologies to poor people, so the elements of appropriate technology transfer – introducing an exogenous technology – are described. The view expressed here is that this introduction should only take place once indigenous technologies have been carefully evaluated and their merits noted. If there is agreement among the stakeholders that there is something to be learned from exogenous technology, it can be introduced. The technology then has to be evaluated and adapted for the new context. Chapter 4 outlines why this should be done using participatory methods, and how local people's confidence and skills to experiment may be enhanced by so doing.

The main part of the book in Part II describes the course of ITDG's brickmaking project between 1996 and 2000, emphasizing how brickmakers and technologists all benefited by carrying out research work together in the field. Why Peru and Ecuador were chosen for this project is the subject of Chapter 5, and Chapter 6 outlines the initial success in Zimbabwe with the coal-fired clamp prior to the start of the project in Peru.

The coal-fired clamp was not the immediate success in Peru that it had been with brickmakers in Zimbabwe, but as other technology options were sought, the need for a methodology to compare the fuel efficiency of different brick-firing methods became clear. A standard method was developed (Chapter 7), and this method, which has been used by other organizations elsewhere in the world, can be regarded as one of the most significant outcomes of the project. Chapter 8 describes how the project progressed in Peru and Ecuador, from initial difficulties with a kiln that would not burn to a flowering of trials of different fuels and kiln designs. The account describes the process by which brickmakers in different communities agreed to work with the project, began to participate in trials, visited other areas to observe new methods, became confident in new methods and were finally able to demonstrate them to others. Chapter 9 outlines the project outcomes as they were identified by the project evaluation in 2000.

Making participation work is not always a straightforward matter. The issues that arise during a project are many, various and often unpredictable, as is illustrated in Chapter 8. The structure of a project therefore needs to be flexible in order to accommodate organic development. Accepting this, Chapter 10 provides some guidelines, but not prescriptive rules, on how to run a technology transfer project with the participation of the local clients. Details of the main events of the project are given in the Project Time Frame in Appendix 1. Examples of completed energy monitoring forms are given in Appendix 2, together with a blank form that can be photocopied and used by readers.

The purpose of this book is to encourage others to develop an approach to working with small-scale producers that encourages participation, rather than imposing more rigid approaches that might be suitable for technology transfer and development in formal sector industries. The main body of the text is illuminated by the inclusion of interviews, anecdotes and articles gleaned from professionals and practitioners working in the field. The book is intended primarily for project managers, decision makers and development workers, particularly fieldworkers engaging with small-scale producers. It will also be of interest to brickmakers, small-enterprise support organizations, training organizations, those involved in technology transfer, students, scientists and technologists – especially those interested in fuels and energy.

PART I
APPROPRIATE TECHNOLOGY TRANSFER

2
BRICKMAKING – THE TECHNOLOGY AND THE PRODUCT

IN THIS SHORT CHAPTER, the nature of a fired-clay brick itself is explained, while the technology of making fired-clay bricks is introduced.

What *exactly* is a fired-clay brick?

The basic ingredient of bricks is clay. The clay selected must have certain properties: it must be plastic when mixed with water so it can be moulded to shape; it must possess enough tensile strength (strong cohesive forces) to keep that shape; and the clay particles must fuse together when fired. Clays are essentially compounds of silica and alumina. Calcareous clays contain calcium carbonate and tend to burn to a yellow or cream colour. Non-calcareous clays typically contain feldspar and iron oxide. Depending largely on the oxide content, these bricks burn to brown, pink or the proverbial brick red. Calcium carbonate, feldspar, micas and iron oxides act as fluxes when bricks are fired; that is, they ease the process of vitrification – the formation of the glassy phase – helping it melt and spread.

Fired bricks consist of crystalline phases held together by a glassy phase based on silica. The glassy phase forms when the clay is heated to a temperature typically between 900 and 1200°C. It melts to spread around the inert crystalline phases, bonding them together. Fired compounds like this are called vitreous ceramics. The temperature at which bricks are fired is critical. If it is too low, bonding is poor, resulting in a weak product. If it is too high and too much glass forms, the bricks slump or melt.

For most applications bricks do not need to be completely vitrified. It is acceptable if the more refractory particles (those resistant to change by heat) are united *enough* to give bricks sufficient strength and resistance to erosion to fulfil their function. Vitrification is seldom achieved in artisanal brick firing around the world.

A well-known technology?

Bricks are one of the oldest known and most widely used building materials in the world. The history books tell us that clay bricks were first made in Sumeria in the Middle East almost 5000 years ago. Today, brickmaking is a familiar technology in most countries and fired-clay bricks are one of the main components of construction in buildings of all types and sizes. They are used extensively in everything from basic housing to commercial and industrial developments. Nevertheless, in the majority of developing countries, small-scale brickmakers remain unfamiliar with efficient and cost-effective production techniques. The bricks they produce are often of poor quality and fail to meet the standards required by the commercial market. The 'farm bricks' of Zimbabwe, for example, though adequate for single-storey buildings, do not come up to the standards relating to strength, water absorption and erodability required for urban buildings (see Chapter 6 and Box 6.1)

In some instances, although soil and energy resources are available, a lack of knowledge prevents poor people in developing countries from exploiting the opportunity to make fired-clay bricks. For these people, brickmaking could offer a much-needed source of income. Furthermore, bricks could be a strong and viable material for building their own homes. It is low-income, poorly housed people who suffer most when natural disasters such as earthquakes and floods occur. The appropriate use of bricks could save many homes from damage and destruction in such events.

The stages of brickmaking

Although a great deal of knowledge has been gained and many technical innovations made, the production of bricks still follows the same basic principles and procedures as it did 5000 years ago. The process involves clay extraction, clay preparation, moulding, drying and firing. The first four of these stages are only discussed in passing in this book since the main aim of the project was to disseminate technologies that improve firing efficiency. Apart from the firing process, however, factors that also have a bearing on the eventual quality of the brick include:

- The quality or nature of the soil used.
- The attention paid to the soil preparation processes. These processes include sieving out stones, mixing the clay – possibly adding sand to limit drying shrinkage – and tempering (soaking the soil for an optimum period, dependent on soil type, to encourage homogenization). Some brickmakers employ weathering, whereby the soil is dug and stacked in advance of making bricks – sometimes for a matter of years. This probably has the same function as tempering.
- The forming process employed. In sand-moulding, sand is used as a releasing agent in the mould. Bricks are formed with drier clay than is the case with slop-moulding; they retain their shape, and the final product is of a higher quality. In the slop moulding process, water is added to clay to form a wet – semi-liquid – mixture and this mixture is packed by hand into a brick mould.
- Attention to drying (green bricks can distort or crack if the drying process is performed poorly).

Much innovation has focused on the firing stage, particularly on the design of clamps or kilns. A clamp is a stack of bricks set up for firing that does not have a permanent structure. Unlike a clamp, a kiln is usually is taken to refer to a permanent structure for firing bricks. In

BRICKMAKING – THE TECHNOLOGY AND THE PRODUCT 15

Photo 2.1 Bricks being sand-moulded in La Huaca, Peru: the mould is first lined with sand, then clay is pressed into the mould (shown), and finally the brick is released from the mould and laid out to dry

Lucky Lowe/ITDG

many areas of Peru, the Scotch kiln was the traditional technology, which had the advantage over brick clamps of being relatively easy to load and unload, and of having better thermal insulation.

Clamps and Scotch kilns operate on a natural draught. That is, air is aspirated through the openings of the firing tunnels, induced by the buoyancy forces of the hot combustion gases while the kiln is open to ambient air. The operation of the kiln during firing, therefore, requires the operator to control or mitigate the factors which influence draught, such as wind speed and direction, kiln height, dampers in the firing tunnels, inter-brick spacing inside the kiln, the moisture content of the bricks, etc.

Brick clamps

Brick clamps can range in size from clamps small enough to fire 5000 bricks to those constructed to fire 100 000 bricks. They may be built into a variety of shapes, but the simplest shape is rectangular in plan, and is built on a platform of pre-fired brick. A typical, basic fuelwood clamp is built up straight for about 10 layers of bricks, then gradually tapers for 15 more layers, ending with a flat layer. The tapering sides give the clamp some stability, which is necessary during the heating and cooling phases when the bricks inside expand and contract.

Fuel is introduced through the firing tunnels, which run from one side of the clamp to the other at regular intervals. These tunnels are usually located along the bottom level of the kiln. Firing tunnels are typically 0.5 m wide by 1 m tall, tapering to a point at the top. Wood may be loaded into the tunnels at the start of the firing, and is fed in when necessary for the duration of the firing.

If the clamp is to be ignited by an oil-fired burner (a technology that was taken up enthusiastically by some of the brickmakers in Peru and Ecuador during the course of the project – see Chapter 8), the firing tunnel is lower, or it may be divided halfway down by a layer of bricks supporting coal above the layer, with room below for the pipework of the burners. The burner can be moved down the length of the clamp to one tunnel after another until all the fuel is ignited. As the firing progresses, too much air causes the clamp to cool down, and brickmakers control the air supply by partially blocking off the fuel tunnels at various stages.

Coal-fired clamps of the kind that became popular in Zimbabwe have layers of coal distributed at intervals all the way up the clamp, between the layers of unfired (or 'green') bricks (see Figure 3.1 and Chapter 6). Unlike bottom-fired brick clamps, it is not necessary to include large firing tunnels: kindling can be put into the gaps in the bottom layer of coal, and these can be sealed up once the kiln has started to burn. It was found during the course of the project in Peru and Ecuador that placing coal throughout the clamp, and especially in the 'cool spots' in corners, produced a more even firing, with fewer under-fired bricks. The insulation of brick clamps, and thus their energy efficiency, is greatly improved by plastering over the sides with a layer of mud, at least 4 cm thick, which is called scoving. This plaster also prevents more air than is necessary from entering the kiln. The top is not covered initially, to allow the residual moisture to leave the kiln as quickly as possible.

In a clamp, bricks are always fired in batches. Some types of large, permanent kiln, such as the Hoffmann kiln, however, are operated continuously – fired bricks being removed while new, 'green' bricks are loaded in or 'charged'. The advantage of this is that the waste heat from the kiln is used to drive off residual moisture from the green bricks.

The quality of the final brick depends to a large degree on the firing process, and how thorough this has been. Under-fired bricks are little more than soil blocks, prone to

Photo 2.2 The traditional wood-fired Scotch kiln of Peru has a permanent structure into which the bricks are loaded and unloaded

Lucky Lowe/ITDG

Photo 2.3 The Scotch kiln in Peru showing the firing tunnels with the oil-burner in place

Photo 2.4 The coal-fired clamp in Zimbabwe has fuel distributed in layers at regular intervals throughout the clamp

Lucky Lowe/ITDG

Box 2.1   Brickmaking around the world

Improvements to kilns and their operation enable small- and medium-sized enterprises (SMEs) to achieve energy efficiencies of at least 40 per cent. It is possible to substitute fuelwood by coal, charcoal or oil. The use of residues (industrial and agricultural wastes) as fuel has great potential in brickmaking. Savings on fuel are also crucial in enhancing the economic sustainability of SMEs. Better firing, combined with other improvements in production, can raise the quality of the end product and increase access to markets and viability.

The introduction of radically different technologies, such as vertical shaft brick kilns (VSBKs), is more likely to fail due to non-technical problems within SMEs (see Box 4.1). Such technologies require a more intensive support package. They usually require major changes to traditional practice, including a very different approach to operational and commercial management.

The participation of producers is an essential factor in successful technology development or adaptation. As well as work in Zimbabwe and Latin America, ITDG is working with brickmakers in Sudan and Sri Lanka.

In Sudan, the brick industry consumes over half of the fuelwood nationwide. In the east, where ITDG works, suitable wood has to be brought from the Blue Nile region, far to the south. Bricks are usually made by slop moulding, which is fast, but produces poor-quality bricks. Some cow dung is mixed in with the clay and bricks are fired in large clamps. ITDG's intervention has focused mainly on better moulding and fuel substitution. Slop moulding was replaced by sand moulding. The resulting bricks are better, much in demand, and fetch a higher price.

Research into fuels has followed different strands. Using an oil-burner is one promising option. The increased use of cow dung, embodied in bricks, as a fuel is also of interest. Cow dung is considered a good additive because it increases plasticity and reduces breakage. Using sand-moulding and increasing the proportion of cow dung used enabled ITDG to reduce the energy share of production costs from around 53 to 36 per cent, whilst saving 44 per cent on wood.

Experiments were also carried out in Sudan using bagasse, a residue which is widely available from sugar factories, as a fuel. It can be mixed into the clay or burnt as briquettes. One test clamp used 114 g of bagasse per brick, replacing cow dung, which came to 12.4 per cent of overall costs. The amount of wood used was 6827 kg, representing 15.1 per cent of total costs. The overall energy use was 2.3 MJ/kg fired brick.

In Sri Lanka there are more than 5000 brickworks, producing over 500 million bricks per year; 85 per cent of this production is by SMEs. There is widespread use of both clamps and Scotch kilns. The brick and tile industry consumes in excess of 150 000 tonnes of woodfuel per year, half of which is rubberwood. Some producers have begun using sawmill offcuts and a few use rice husks. ITDG's main objective is to investigate the increased use of residues, such as sawdust, coir dust and rice husks, as fuelwood substitutes.

Source: Schilderman, 1998

Layer stacking to produce the clamp structure

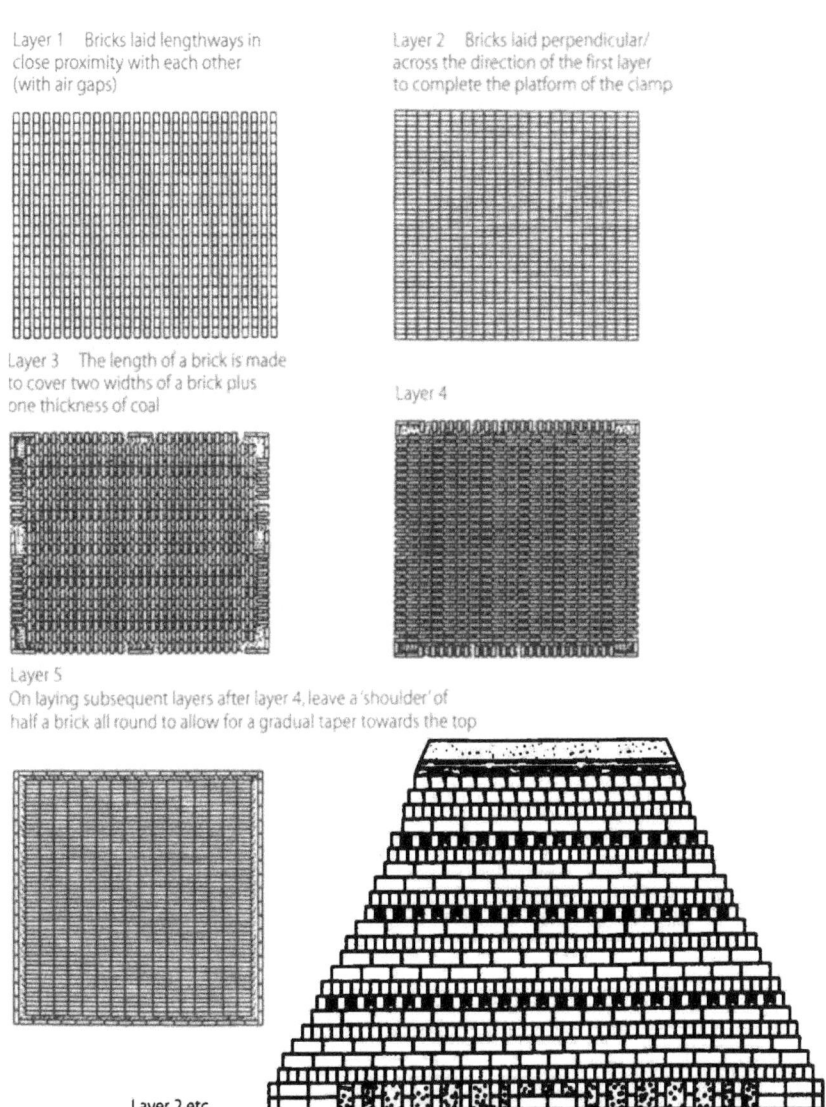

Figure 2.1 The coal-fired clamp – a technology which transferred successfully to Zimbabwe

crumbling and rain erosion, while over-fired bricks are usually distorted or cracked. Exactly how much fuel to use is for most small-scale brickmakers more a matter of judgement and experience rather than measurement (or, as one fieldworker put it, more like cooking than science, see Box 7.5); and whether the bricks have been fired sufficiently can only be determined once the clamp has finally cooled down and the bricks have been extracted.

Research on clamps and kilns has naturally focused on controlling the burning process, the aim being to produce bricks using less fuel and thus to reduce costs. Besides being a major cost, fuel is often in short supply. Also, global awareness of the negative environmental impact of all burning processes has been raised.

Generally, brickmakers determine the most suitable burning process for their particular situation. But the process selected may *not* be the most efficient or environmentally friendly – the incentive may not be there. When fuelwood is freely available there is no pressure for energy efficiency. In some cases, the process may be such an established and accepted one that no one thinks to challenge it. Often information is simply not available. Yet, in many instances, a number of relatively minor innovations that would save fuel and reduce environmental damage could be adopted, drawing on the experience of other brickmakers around the world.

3
APPROPRIATE TECHNOLOGY AND ITS TRANSFER

THIS CHAPTER SEEKS to define appropriate technology and technology transfer, and to identify the key elements in a process of technology transfer. The significance of appropriate technology in the debate about globalization and dependency is raised, and a working definition of appropriate technology is presented. Moving on to technology transfer, differences in the process as implemented by the commercial and non-government or development sectors are identified. The key questions raised are how is technology transferred and when can its transfer be said to be successful? The need to adapt technology for use in different settings is discussed. Two case studies are presented which highlight the critical questions that confront organizations contemplating a technology transfer project.

Appropriate technology – in search of a working definition

Appropriate technology is inevitably subject to a number of definitions. What is appropriate and what are the terms by which it is defined? Without getting too deeply embroiled in the on-going debate, it is worth highlighting some of the key arguments. There is some agreement that an appropriate technology is one that makes the best use of available resources. In most developing countries this will mean labour- rather than capital-intensive technologies. This is because unemployment or under-employment is very high in most developing countries. Meanwhile, capital – particularly hard currency – is scarce.

Some observers believe in addition, however, that for a technology to be appropriate it must benefit specifically low-income producers, who are those largely left out of the mainstream process of economic development (Stewart, 1978). There are also those who believe that an appropriate technology has to produce an appropriate product and benefit low-income consumers. In these terms, appropriate technology must produce inexpensive goods for poor people.

Under this definition, what constitutes appropriate technology is considerably restricted by socio-economic and moral considerations. Not only must it be labour- rather than capital-intensive, it must benefit both low-income producers and consumers. As such, appropriate technology would be confined in both scale and scope; it would be hidebound – unable to develop and change as market opportunities arise and according to the creativity of producers.

This constraining of appropriate technology on both the supply and demand sides merits further discussion. The premise is that high-quality products are in demand from high-income consumers, and low-income consumers are in the market for labour-intensive, low-skill, low-quality products. But high-income consumers may want to choose craft-intensive, high-quality products as well as highly mechanized, high-quality products. Meanwhile, modern technology – mass production – can be harnessed to produce goods for low-income consumers. The relationships between choice of product and choice of technique, and between income levels and income distribution, are obviously complex.

Improved craft or manual production may open up a luxury market to small-scale producers. They would then, however, be meeting the 'needs' of high-income consumers, and some would judge their technology therefore to be inappropriate. It is suggested that the idea that small-scale producers have to supply low-income consumers is nonsensical. Such a constraint would confine them to remaining poor themselves. It would maintain the inequitable status quo.

Brickmakers in certain locations in Peru, for example, were keen to diversify their product range in order to supply more lucrative markets. They had saturated the market for low-cost, low-quality bricks and were anyway limited in the margin they could make on such products. Producing better-quality bricks and manufacturing roof tiles were both within the scope of their natural resource and technical potential. Should they then be confined to producing low-cost bricks for exclusively low-income consumers – a commercial dead end?

Within ITDG's Building Materials and Shelter Programme, this question of whether interventions in activities such as brick production should benefit poor consumers as well as poor producers is a frequent topic for debate. The working conclusion is that benefiting poor producers is justified in itself. The unfeasibility of choosing only those projects in which poor consumers benefit is well illustrated by the programme to assist small-scale gold producers – how could they produce gold for the poor consumer?

At the other extreme, some technocrats might define an appropriate technology as a set of techniques that make optimum use of available resources in a given environment. In these terms, nuclear power would be an appropriate technology for a nation with no oil reserves but having uranium deposits and capital to invest.

Some political analysts even argue that appropriate technology is not an alternative for developing countries. They believe it is doomed to failure. The argument is that the whole global system is geared to the maintenance of inappropriate technology. There is a lack of money going into appropriate technology because rich countries control the research and development. In these terms, the effective pursuit of an appropriate technology would threaten interests in the advanced countries as well as those of élites in under-developed countries who are currently benefiting from the use of advanced-country technology.

The continued use of advanced-country technology is at the heart of the dependence of the poor countries (see Box 3.1). It permits, indeed necessitates, the continued sale of technologies, goods and managerial services to poor countries on terms favourable to rich countries. The transfer of advanced-country technology maintains the status quo, benefiting only the élites in both the exporter and importer nations. A move away from trade with advanced countries towards trade with other developing countries, it is claimed, is a necessary condition for the pursuit of a more appropriate technology.

Appropriate technology – software as well as hardware

ITDG's mission is to build the technical skills of poor people in developing countries, enabling them to improve their lives and those of future generations. With that in mind, ITDG is critically interested in a definition of appropriate technology that is positive, moral and practicable. Dr E.F. Schumacher, the founding father of ITDG, held that an appropriate

> **Box 3.1 Ghandi's critique of industrial society**
>
> A great measure of world poverty today, and African poverty in particular, is due to the continuing dependence on foreign markets for manufactured goods, which undermines domestic production and dams up domestic skills, apart from piling up unmanageable foreign debt. Ghandi's insistence on self-sufficiency is a basic economic principle that, if followed today, could contribute significantly to alleviating Third World poverty and stimulating development.
>
> Ghandi remains today the only complete critique of advanced industrial society. Others have criticized its totalitarianism but not its productive apparatus. He is not against science and technology, but he places priority on the right to work and opposes mechanization to the extent that it usurps this right. Large-scale machinery, he holds, concentrates wealth in the hands of one man who tyrannizes the rest. He favours the small machine; he seeks to keep the individual in control of his tools, to maintain an interdependent love relationship between the two, as a cricketer with his bat or Krishna with his flute. Above all, he seeks to liberate the individual from his alienation to the machine and restore morality to the productive process.
>
> As we find ourselves in jobless economies, societies in which small minorities consume while the masses starve, we find ourselves forced to rethink the rationale of our current globalization, and to ponder the Ghandian alternative.
>
> *Source:* Mandela, 1999

technology was simple, small-scale, low-cost and non-violent (Schumacher, 1974). Over time, these principles have been expanded upon. The trend is away from definitions of technology as machinery towards greater recognition of the software involved. Technology is more than the use of tools, machines, materials and sources of power to make work easier and more productive. Technology is not just about tools, but a kind of tool-using behaviour: a set of methods for producing specific goods and not merely the machines to make them with. Technology is composed of knowledge and skills, organization and product, as part of a dynamic system. Appropriate technology may still ideally be simple, small-scale, low-cost and non-violent, but these qualities no longer define it. In other words, both software and hardware issues need to be considered.

In the past, technological interventions have often been based on the assumed technological superiority of 'foreign experts' – mainly men – from 'advanced countries'. Technologists brought pre-determined solutions to 'people in need' in developing countries. This approach served to undermine the confidence of the indigenous people, especially women, who found that all too often their existing technology and skills were undervalued or completely ignored. The new approach is to recognize and investigate existing knowledge and skills. Building on these in co-operation with the local community has been demonstrated, through experience, to be essential if sustainable technological capability is to be developed.

The definition of an appropriate technology should therefore be one that builds on indigenous knowledge and skills with the participation of the local community. In particular, the knowledge and skills of women should be recognized. This approach moves away from an 'outsider' definition of appropriate technology and allows communities to define their needs and what is appropriate for them. What an external agency can bring to the process is access to information: the aim is to facilitate the transfer of that information based on an in-depth knowledge of people's defined needs. The agency assists in transferring and adapting technologies to each new context, but this must be achieved without creating dependence on external inputs.

The technology transfer organization brings to the process of technology development specialist knowledge, in the form of skilled people, as well as the ability to lobby for change, to be an advocate and to apply effective leverage on behalf of the communities it assists.

Bearing all this in mind, an appropriate technology may now be defined as one that makes the best use of available resources, assuming 'best' includes environmentally friendly and gender-sensitive, and that it benefits poor and marginalized people. Furthermore, in most developing countries in most cases, appropriate technology *will* mean labour- rather than capital-intensive technologies (see Box 3.2). Of course, products manufactured with appropriate technology must be price and quality competitive. But in addition, an appropriate technology can only qualify as such if it succeeds.

Defining technology transfer

Turning to the brickmaking technology transfer project in Peru and Ecuador, one of the first tasks the project team completed was a review of the theory and experience of technology transfer. Much of the information gleaned dealt with the commercial and industrial sector. That is, to instances where a company transfers a proven technology to a different setting, frequently in a different country. The company will, in all probability, have a profit motive and a long-term stake of some kind in the new enterprise. Hence, much of the concern is with legislation: patent agreements, technical assistance agreements, licensing agreements, joint ventures, tax incentives and so on.

Technology transfer in the non-government, small-scale sector is usually a very different process. Resources are much more limited and a non-profit making organization, such as an NGO, will often play a facilitating role. An NGO will, almost certainly, wish to make the project self-sustaining: it will want to cease or at least minimize its involvement at a certain, designated point. Nevertheless, many of the observations, principles and practices involved in technology transfer will be common to all sectors. The experience of different types of organizations is therefore worthy of consideration. It is difficult to separate references to technology transfer from those on technology choice, transformation of technology or appropriate technology. There is evidently an overlap between these topics.

There are a number of opinions as to what technology transfer is and what it involves. As was discussed earlier, technology should be regarded as more than just tools, and should also include tool-using behaviour. In other words, technology is a package of scientific, technical, industrial, administrative, financial and marketing knowledge (Manser and Webly, 1979). Logically, technology transfer would then be the communication and adoption of such knowledge. Some definitions, however, stress the need not only to supply such a package but, more importantly, stimulate the development of local capabilities and institutions. Transfer is judged a success when an indigenous capacity to generate technology has been developed.

When a technology is transferred from a developed to a developing country there will almost inevitably be the need to adapt it to suit local conditions. Indeed, this is the case in most instances of technology transfer. For a developing country, though, the need is not for a

Box 3.2 Scales of production in Ghanaian brickmaking

A recent study of brickmaking in Ghana was commissioned to investigate why the industry as a whole was doing so poorly, and why many large factories had closed down. A survey examined the technology used, and economic constraints experienced by a sample of large-, medium- and small-scale brickmaking factories. From the results of the survey, the following are the main stages in brickmaking and the technology requirements at each stage:

- *Clay winning.* This is carried out using clay digging machinery, such as grab excavators, by large-scale factories; and by labourers using pick axes and wheelbarrows for small-scale factories.
- *Clay preparation and mixing.* Screens, crushers and disintegrators are used in large factories; some small operators use machinery, or else they mix the clay manually.
- *Moulding process.* Hand-moulding processes are common in small-scale factories; in medium- and large-scale factories this is done by machine.
- *Drying of green bricks.* Small-scale factories dry bricks by stacking them on racks around which the air may circulate. Large-scale factories usually build drying sheds where hot air from the kiln can circulate around the green bricks.
- *The setting up and firing of kilns.* This is done manually by small-scale operators, and is mechanized in large-scale operations. Small-scale factories use fuelwood and charcoal briquettes for firing; they are finding it increasingly difficult to get sufficient supplies of wood legally. Large- and medium-scale factories use petroleum products for firing; these are currently in good supply in the country.

All of the machinery investigated by the survey was imported, except for a pugmill at one factory and a hand-operated tile-making machine. During the installation and trial of the imported machinery, costly foreign labour was hired for some time: sometimes for several months. This was not necessary for small-scale factories.

Economics of production. The unit production cost of standard, solid bricks from the medium-scale factories is nearly twice that of the small-scale ones. This is partly because of the higher capital costs involved in purchasing the machinery for large-scale factories, which, since the machinery is imported, is exacerbated by the falling value of the cedi against other foreign currencies. Bank interest rates are high, and working capital for purchasing fuel is in short supply. Large-scale factories therefore often stand idle (one for as much as 25 per cent of the time); medium-scale factories also only achieved production capacities of about 50 per cent. Furthermore, in the event of machinery breaking down, spare parts were either unavailable or very expensive.

Finally, the cost of transporting bricks from the factories is fairly high, and beyond a certain distance it is not economical to transport the product. All of these constraints to production affect large-scale mechanized factories much more than small-scale labour-intensive ones.

Since the date of the survey, the majority of the large-scale brickmaking factories have closed down; some of the semi-mechanized factories are operating, and

> most of the small-scale factories are thriving. It was clear from the results of the survey that in a country like Ghana, where economic conditions are unstable, technologies adopted for brickmaking should either be small-scale and labour-intensive, or medium-scale and semi-mechanized.
>
> *Source:* Hammond, 1997

developed-country technology but for technology that is appropriate to the particular goals and conditions of that country, which generally include capital scarcity and an abundance of labour.

Technology transfer issues are therefore different in the case of the small-scale sector in developing countries. There are at least two important reasons for this. First, developed countries do not have a monopoly on technologies that could be most appropriate for developing countries. Technologies appropriate to developing countries may take the form of older technologies from so-called advanced countries; they may be 'intermediate technologies'. When this is the case, many of the legislative issues such as intellectual property rights, licensing and so forth do not apply.

Second, private companies in the advanced countries usually have no incentive to undertake research and development to meet the specific requirements of developing countries. It may well be to the advantage of these companies to encourage developing countries to buy in technology and not attempt to develop an indigenous capacity. By retaining the technology they foster dependency.

It seems reasonable to propose that technology has been transferred when the local workforce is able to take charge of the imported technology and to do so efficiently. There is general agreement in the NGO sector that the aim should not be to transfer technology *en bloc* from advanced countries but to create a technical capacity in developing countries themselves. By the end of the brickmaking project in Peru and Ecuador, for example, no single new fuel had been adopted from abroad, but brickmakers in several communities were experimenting with a whole range of different fuels in different combinations, and had experience in measuring the energy efficiency of these different methods (see Chapter 8).

Some observers state that technology transfer has occurred when the technology spreads to other productive units in the recipient economy, either through active dissemination or by assimilation and imitation (Santikarn, 1981). Another view is that successful projects are those that maximize community involvement in basic needs production, using appropriate technology to avoid over-dependence on expatriate finance or management, technology or skills (Browne, 1982).

However complex the definition becomes, the crux of technological transformation for small-scale producers in developing countries lies in their capacity to improve on traditional technologies (UN, 1986). Imported technologies must be carefully chosen, adapted and utilized. Developing countries should strive to innovate or generate new technologies in line with their overall development needs. It is worth remembering that technology transfer is not an end in itself. The overall objective should be economic development and growth (Stewart and Nihei, 1987). Or, in more human terms, poverty reduction through the creation of jobs and income.

Elements of technology transfer

Local research and development. When it comes to the elements that make up a successful programme of technology transfer, one recurring theme is the desirability of establishing the capacity for local research and development. Sometimes this is achieved by setting up locally managed and operated research centres. Research is potentially one of the most fruitful areas of technological co-operation. Commercial sector companies have recognized the merits of using local scientific and technical services whenever possible. Often they seek to use local staff and to extend relations with universities and technical colleges. Such companies tend to use local academics as consultants and to send their own staff into universities and colleges. In short, carrying out research and development locally, employing local resources and encouraging intellectual cross-fertilization are highly effective elements in successful technology transfer.

Training or capacity building is an essential element. Commonly, there is a scarcity of technical and managerial skills in developing countries. To facilitate the adoption of the new technologies, assistance is needed to train technologists, supervisory personnel and managers. This builds the capacity of the institutions that they work for. It is accepted practice for some commercial companies to train more technicians and managers than their project actually needs. The idea is not only to anticipate growth and staff migration, but also to contribute to a reservoir of skills that will 'trickle down' and act as a catalyst for more general economic development.

Appropriate or intermediate technologies are frequently mentioned as the right choice for developing countries, especially amongst the NGO sector. Basically, intermediate technologies lie between the traditional and the fully automated. As mentioned in the previous section, it is evident that developing countries generally have different resources – more labour as opposed to capital (the difficulties of installing and maintaining high capital-cost machinery for large-scale brick factories in Ghana is described in Box 3.2). It is quite widely recognized that a technology will at least need to be modified to suit the local context. And not only should the technology be appropriate for a developing country, but so should the final product. Unfortunately, these considerations are not always taken into account by commercial investors with perhaps no developmental agenda or broader commitment to economic development in the 'target' country.

The onus is often placed on developing countries to choose technologies appropriate to their socio-economic conditions. In the real world, however, the government of a developing country is unlikely to turn down the offer of, say, a car manufacturer to invest in a local assembly plant – no matter how inappropriate the technology it intends using. Most governments simply could not afford to say no; they would not have the bargaining power to influence the technology chosen. Typically, developing countries are crying out for investment under any conditions. Manufacturers know this. So, offering any kind of resistance will probably result only in the assembly plant being built in country B rather than country A.

Morally, the onus should also be on encouraging manufacturers to recognize the developmental needs of the countries where they work. There is a similar onus on NGOs when assisting the small-scale sector (though Box 3.3 cites an example of an NGO that was prepared to exploit its clients). For example, what poverty-stricken brickmaker is going to turn down the offer of help or have the knowledge in the first instance to question the appropriateness of that help? Thus, the NGO will inevitably be responsible for making some initial critical decisions concerning its intervention.

> **Box 3.3** The moral imperative
>
> The morality of an NGO cannot be taken for granted. In Zimbabwe, for example, an NGO began a brickmaking project with a group solely for the purpose of making bricks to build their own offices. They did not tell the group their agenda. When they had enough bricks to build their offices they simply withdrew their technical, managerial and financial support, abandoning the project and the people. The fledgling enterprise dwindled painfully out of business. More than 20 people and their families lost their source of income. Expectations had been raised and false promises made. The NGO's fieldworkers were scandalized, but powerless to do anything if they wished to keep their jobs.
>
> Everyone has the moral imperative to examine the organization he or she is working for. Simply because an organization is nominally a charity or an NGO does not make it incorruptible or guileless. 'Righteous' NGOs should operate in a transparent way, encouraging their allies to question their agenda, attitudes and working practices, to talk to other allies and to examine their track record.

Establishing marketing facilities and paying attention to marketing is another critical aspect of facilitating technology transfer (van Ginneken and Baron, 1984). This is particularly important where it is intended to introduce a new product. It will be vital if the product has competitors. But marketing should not be ignored even if there are only minor changes to the product. For example, there would be no point whatsoever in encouraging and assisting small-scale brickmakers to improve the quality of their bricks unless there was a market prepared to compensate them for their efforts: customers who wanted better bricks *and* who were prepared to pay more for them. For the brickmakers in La Huaca, Peru, for example, savings derived from using new, more fuel-efficient technologies were almost swallowed up as the demand for low-cost bricks fell and prices dropped. The local brickmakers' association knew that they had to reach new markets for their higher-quality bricks.

Capital. The scarcity of capital in developing countries has to be addressed at all scales of technology transfer. If an investment is essential, especially if it is to buy imported hardware, then funding has to be available. In the case of the small-scale sector this is unlikely. The most appropriate technologies may, therefore, be those that are generated locally and do not require significant investment. In some cases, however, capital will inevitably be required, perhaps in hard currency. The facilitating agency has to plan for this eventuality and find the resources to make grants or loans. In the case of loans, they will also have to plan how they will be repaid and what will happen in the event of a producer defaulting. This is a tricky issue that has brought many projects to grief. Successful means of advancing credit to small-scale businesses have, however, evolved. In Bangladesh, for example, there is a successful history with micro-credit schemes (UNCHS, 1989; Rutherford, 1995; Khandker, 1998; Versluyen, 1999), for example, the Grameen Bank.

The dissemination of information is essential. Whether this is information about the technology or the product is immaterial. The audience – the potential beneficiaries or customers – must be informed before they can be expected to make a choice. An NGO working as a facilitator or technology promoter in the small-scale sector will perhaps most often be working with limited resources and therefore a small group of producers. It will be impossible to reach a national let alone an international audience directly. Thus, in order that the technology should be more widely adopted, it is essential to disseminate information of the appropriate content. Appropriate because an NGO may have different audiences in mind. On the one hand, small-scale producers will need practical, affordable and actionable information. Meanwhile, policy and decision makers, potential customers and other NGOs will require substantially different information.

Alliances or partnerships are crucial. An NGO working in a developing country will need local allies appropriate to the nature and scope of the projects it is engaged in. Allies might include universities, other NGOs, producer associations, government departments or local authorities. In many cases these allies will require infrastructural support.

On-going learning. Project promoters should understand the dynamic nature of technology. On-going learning is vital to the success of all ventures, particularly perhaps small-scale industries (Smillie, 1986).

Monitoring and evaluation is essential to assess the efficacy of the process of technology transfer and facilitate on-going learning. Constant monitoring and regular evaluation helps identify problems and allows them to be addressed promptly. If the process is to be improved upon or duplicated, the experience must be well recorded, analysed and documented.

Effective communication underpins all the elements listed above, and this cannot be left to chance. Mechanisms – meetings, seminars, reporting – should be put in place to ensure clarity, common purpose and transparency. Speaking the same language – defining and agreeing terms – is essential and worth spending time on (see Box 3.4). For example, Dr Alfredo Barriga commented on the necessity of taking time to check that brickmakers and fieldworkers mean the same thing by a term like 'quality': for the fieldworkers it implied strength and water resistance, but for the local brickmakers it meant simply 'transportability' (see Box 8.8).

In many cultures there is the tradition of holding 'talking shops'. In Nguni culture in southern Africa, for example, there is the indaba, which is held in a designated house or under an indaba tree. It may be possible to integrate traditional forums into projects – or rather, integrate projects into traditions.

Other factors in the process

Experience dictates that a number of other factors have to be considered. The following list is certainly not presented as comprehensive. Each project or programme will doubtless face unique issues. These should be identified and addressed at the earliest possible stage. In certain cases, the factors mentioned here might well emerge as critical to successful technology transfer.

Quality. In the instance of a company making the choice to manufacture its product in a new location, attention will have to be paid to the maintenance of quality. As far as possible, our hypothetical car manufacturer will want models assembled in country B to be the same as those from country C. In practice, this quality control is often difficult to achieve. In southern Africa, for example, it is recognized or at least generally perceived that vehicles assembled in South Africa are superior to those from Zimbabwe. There again, vehicles imported direct from Japan are thought to be better still.

Health and safety. Whatever the product, no facilitator of technology transfer – be it a commercial company or an NGO – will wish to be associated with an inferior, dangerous or environmentally harmful product. An NGO will have to pay attention to health and safety in

> **Box 3.4   Communication is the key**
>
> Anybody who tries to introduce technology into a society has to start off by getting the language right. What separates scientists and engineers from the general public is a different vocabulary. There is no point using words like dioxin or ozone in the newspapers if it does not mean anything to the reader. Unless you can use words which really mean something to your audience, you cannot begin to communicate.
>
> *Source:* Wolf (1994)

the sector it is assisting. In some cases, this could generate a conflict of interests with employers unused to such considerations. The issue is not at all straightforward. Should an NGO, for example, insist that the employees of the brickmakers it assists be equipped with steel toe-capped boots? Traditionally barefoot employees would certainly be safer wearing such boots, but the outlay would probably bankrupt a brickmaker working on a very slim profit margin. Well-shod workers could end up with no job to go to.

Legality. This is an additional concern of NGOs often working with the so-called informal sector. Small-scale brickmakers, for instance, typically operate outside 'the system'. They may produce bricks that do not meet materials standards for safety. For certain applications these bricks may be dangerous to build with. Again, brickmakers may well operate outside the tax system and ignore legislation such as minimum wage rates. The NGO has to be careful in defining its associations and alliances. Imagine a house collapsing because it was built with sub-standard bricks produced by brickmakers assisted by an NGO. Few commentators would be interested in the *details* of the assistance. There are many potential pitfalls, but they are integral to the challenge of working with the poorest sectors of the community.

Case studies in technology transfer

In terms of brick and ceramic production, there are two specific case studies on technology transfer to which ITDG project staff in Peru paid particular attention before launching their intervention.

The brick industry and technology in Malaysia

At the time of this study (taken from Lim, 1978), the industry was open to expansion due to increased demand for houses. Firms that used hand-moulding technology and intermittent kilns, as opposed to extruders and continuous tunnel kilns, were almost always very small. Large firms tended to use hand-moulding techniques only to produce 'specials' – bricks shaped to a customer's specific demands. The small firms had low fixed assets, low wage levels, were typically based in rural areas and employed readily available seasonal labour.

The advantages of small-scale operation were that electricity was not required and the enterprises retained flexibility. For example, they shut down for the three months of the monsoon season. Natural drying, employed by the small companies with intermittent kilns, had a negligible capital cost and meant a major saving in fuel costs. The disadvantages were that breakages tended to be higher, the process needed a lot of space and it could not be controlled.

Overall, the scale of production determined technology choice. It could, however, be argued that the opposite was also true – technology choice determined the scale of production. The factors affecting technology choice were listed as:

- access to credit;
- access to information;
- level of education and training of the decision maker in the business;
- extent of foreign ownership;
- fiscal incentives (import duties, etc.);
- labour relations (mechanization might be preferred if the likely alternative is a powerful union);
- location (urban or rural: proximity to and size of market);
- variety of products;
- product quality (the urban market demanding a higher standard);
- extent of clay reserves.

The study concludes that access to technological information is not enough on its own for brickmakers to make the appropriate technology choice. The decision makers in the businesses also had to be in a position to evaluate the information. This depended very much on the level of their education and the type of training they had received. Small firms chose labour-intensive techniques because wages were generally lower and capital more expensive. The existence of a large number of small firms in the brick industry indicated that economies of scale were not significant.

Long term, the study concluded, policies regarding the educational process would be more important than pricing and fiscal policies. In the short term, direct action to facilitate information flows to stimulate diffusion and to train labour were needed. The most urgent requirement was for the systematic provision of information on appropriate technologies. The development of an information system was judged to be essential in influencing the choice of technology.

Village-industry pottery in Ghana

According to this study, the rural production of ceramics had certain common features (Browne, 1981). Clay was obtained without incurring any financial cost; it was 'won' by hand. Potters worked part time, being also involved in farming. They had little formal education or training. No special equipment was employed. Clay was formed by hand without the use of a potter's wheel. Products were fired at low temperatures in rudimentary kilns and there was no glazing. Most potters sold their products to visiting traders, so transport was not a consideration or a cost. Finally, making pots was exclusively a women's activity.

Very interestingly, the study of this industry concluded that no changes could be justified. All options for improvement were considered inappropriate because:

- Clay deposits were likely to be exhausted in 30 years at existing levels of extraction; increased production would hasten depletion and saturate the market.
- Capital investment in wheels and kilns would be expensive and reduce the flexibility to move between pottery and farm work.
- It was suspected that if production evolved to using mechanized wheels and special kilns, the industry would be taken over by men!

New products, such as water coolers and tableware, were judged to be unnecessary because there was no local demand. Forming co-operative agencies was said to be irrelevant since work-place co-operation already existed and buying and selling arrangements were perfectly adequate. This cautionary tale concluded that this indigenously developed technology was already the best available given the local conditions.

These case studies highlighted critical questions for the brickmaking technology transfer project in Peru and Ecuador. Exactly how were brickmakers going to access information on the technology options available to them? Would they be able to evaluate and use this information or would they need education and training? The fieldworkers would have to evaluate constantly the justification of their intervention: had the brickmakers already selected the technology that was best for them given the constraints they worked under?

4
PARTICIPATORY TECHNOLOGY DEVELOPMENT

PARTICIPATORY TECHNOLOGY DEVELOPMENT, or PTD, is defined in this chapter as an approach to deriving new technologies by working with people. The aims are to enhance knowledge and skills in a participatory manner, extending choices and promoting ownership of technology, keeping in mind the objective of community empowerment. Using PTD as a project tool is discussed. Degrees of participation and stages in technology development are proposed, while potential key factors in building technological capability are enumerated. There is an examination of the roles that the 'external change agent' can play in the process of participatory technology development.

Enhancing knowledge, skills and choices

PTD is an approach that ITDG is using in some projects, particularly in its building materials and shelter work. PTD can be useful in facilitating the development of technologies in a particular community and context. As such, it has much in common with technology transfer. Where it differs from technology transfer, however, is that in PTD full 'beneficiary' participation is essential. PTD can increase communities' confidence in and ownership of development interventions. The involvement of the people most affected yields a much better chance of success. Participation in technology development allows for the recognition and extension of existing knowledge. It should contribute to the development of technological capability and have a positive impact on the lives of poor people.

There is a wealth of literature that examines the history and application of PTD, and the interested reader is referred to the Further reading section, 'Participation, training and technology development', at the end of the book. A number of workbooks for PTD-type training of trainers and groups exist, for example Hope and Timmel (1995, 1999). What is covered here is an overview of PTD and a discussion of how it can be applied as a project tool – and how to interpret its impact.

Ownership through participation

In the past, technology development – in the appropriate technology context at least – has been seen as the development and transfer of equipment and skills. Analysing problems, identifying needs and solutions, and technology development were usually, but not always, carried out in consultation with the end-users. Decisions about *which* technologies and skills were needed were, however, usually taken by 'outsiders'. These outsiders were typically technical specialists whose interest was often in 'technology development' rather than 'technology for development'. The transfer of equipment dominated the process and formed the basis of the criteria by which success was measured. The social and managerial aspects of equipment use were often not considered.

Towards the end of the 1980s, appropriate technology organizations started to tackle the poor dissemination rate of the products they had been developing over the previous 15 or 20 years. They began to re-evaluate their roles in relation to the small-scale technology producers and users with whom they worked. They examined the different contexts in which technology development takes place. It was evident that such technology development as had occurred in the countries of the South had neither focused on the poor nor enabled them to become less poor. Not involving the poor in technology development was effectively leaving them behind and increasing their vulnerability.

The concept and processes of PTD were first used within the sustainable agriculture movement (e.g. Murwira et al., 2000), and then they began to be tested in off-farm technology development. The aspiration was that participation in technology development could humanize technical change: it would enable those previously at the 'receiving end' to manage their own process of change. In the short term, a community's objective might be to engineer a particular technical change, but in the long term, they would develop their own processes for managing that change *and* their socio-political environment. Hence, they would begin to challenge the structures that maintained them in poverty and kept them at a disadvantage.

PTD has to deal with two sets of linkages: those between insiders and outsiders, and those between technology and other aspects of people's lives. Insiders are already decision makers in processes of technology use and change. These processes reflect both their knowledge of the context and their priorities. Outsiders have their own knowledge, assumptions, objectives and priorities; these may or may not coincide with those of insiders (see for example the differences in attitude towards the VSBK in Box 4.1). A starting point for PTD is respect by outsiders for local knowledge. They have to be prepared to recognize and challenge their own assumptions about the relative status of their own and other's knowledge, skills and priorities. PTD uses what people know to explain what they do not know. Outsiders also have to recognize that different groups in the community will have different, even conflicting, interests. Insiders, like outsiders, are not a homogeneous group.

People choose to participate in technology development for a variety of reasons. It is often assumed that entrepreneurs are motivated by the short-term profits that can be derived from growth and increased productivity. Security of production, guaranteed markets or production routines compatible with other activities may, however, be more important to them. Also, technology development does not always take place exclusively within an enterprise context. Communities may be motivated by catastrophic changes in their circumstances brought about by natural disasters, economic or environmental crises. In such cases, the predominant motivation may be survival, food security or safety. Participation in technology development means strengthening technical skills. These skills can then be used to identify priorities, make informed choices, and improve status and self-esteem. They can improve a community's chance of survival and help them gain more control over the physical, socio-economic and politico-cultural forces that affect their lives.

PTD as a project tool

PTD is a productive approach to development work: it is not a new 'orthodoxy' that can, by itself, solve development problems. PTD is a management tool, or rather a set of tools.

Box 4.1 The vertical shaft brick kiln in Pakistan – problems with technology transfer

The vertical shaft brick kiln (VSBK) is a technology developed in the 1960s in China, where it is popular among small-scale brickmakers in rural areas. It is remarkable for its high energy efficiency: specific energy values range from 0.8 to 1.4 MJ/kg of brick. This made it a particularly attractive option to consider in Peshawar, Pakistan, where the predominant kiln in use, the Bull's trench kiln (BTK), was relatively fuel inefficient (1.1–4 MJ/kg) and polluting. Between 1993 and 1996, four VSBKs were built in Peshawar, with donor funding, in an attempt to disseminate this technology; however, local brickmakers have remained unwilling to adopt this technology.

This is partly because the VSBK, in the form that it was popularized in China, has a lower capacity (4000–7000 bricks in 24 hours) than the BTK (7000–28 000 bricks in 24 hours), which made it more suitable for rural areas in Pakistan than for adoption by the medium-scale brickmakers operating around Peshawar. The failure to transfer this technology was also due to poor project management, insufficient investment in technical training and a lack of monitoring of the project, which might have picked up and addressed the problems that arose early on in the course of the project.

The introduction of the VSBK. The VSBK consists of a tall, rectangular shaft, into the top of which green bricks and coal are loaded; fired bricks are later removed from the bottom of the shaft. The first VSBK in Peshawar was built in 1991; three engineers were brought from China to construct the kiln and to train local brickmakers. For some reason the engineers constructed an older model of VSBK, which had already been superseded by a model that produced more bricks for less fuel and with a lower initial investment. In addition, the engineers only stayed long enough to build the kiln and run it for a few weeks: the local management was confident that the brickmakers were sufficiently trained and wished to save costs by returning the Chinese engineers as early as possible.

Problems with the first model. The kiln was only run for a few short periods over the next couple of years, and most of the time when it did operate it performed poorly, producing a large proportion of over-fired, under-fired and broken bricks. Continual staff changes meant that there was no accumulation of technical experience with the VSBK. Part of the difficulty was due to differences between Peshawar and China in the quality of clay and of the coal used: the local clays contain more stones and other impurities, and the clay processing does not remove these impurities; similarly the coal is of a lower quality than that used in China. In spite of the obvious difficulties faced in operating this kiln, the technology was not evaluated, and the project went on to build four more kilns around Peshawar in an unsuccessful attempt to disseminate the VSBK. The technology now has a poor reputation among brickmakers in Pakistan.

The need to evaluate current practice. A more participatory approach to technology transfer might have had a better outcome, or at least cut short the wasted expenditure. It would also have involved a more thorough investigation of current practices (the BTK), which had a greater output than the VSBK, making the latter unlikely to be adopted by medium-scale brickmakers around Peshawar. Instead of consulting the brickmakers early on, the brickmakers' association was informed that all the BTKs should be replaced by VSBKs, and this resulted in considerable bad feeling. Scaled-up versions of the VSBK are now being tested, but these are more expensive to construct than the BTK, and the savings in fuel are not great enough to recoup the initial capital investment.

Fortunately this is not the end of the story. Lessons about careful planning and management have been learned from this experience, and when the VSBK was first introduced in India, the project worked well with small-scale brickmakers who were then using brick clamps rather than BTKs. Chinese engineers were brought in for a full six months to build the first kiln and train the Indian brickmakers; and no attempt is being made to disseminate the technology before another kiln has been built and tested in another area. The first kiln is now producing good-quality bricks, with low levels of wastage, high fuel efficiency and low levels of pollution.

Source: Jones, 1997

Photo 4.1 The vertical shaft brick kiln (VSBK): too radical a technology transfer?

Theo Schilderman/ITDG

Development work as a whole has to take place with an understanding of the deeper determinants of technical and social change (Biggs, 1995). PTD focuses on empowerment and adult learning, it involves a process of collective research and development based on a community's needs. It seeks to create an environment in which people can share their ideas and knowledge. This often requires the sensitive support of a facilitator, and this is one of the roles that ITDG is often able to accept.

Thinking in terms of a project may not be compatible with the concept of PTD. While it may help in planning work, it can limit the capacity of organizations to conceptualize technology development as part of the continuous process of community development. People's lives do not take place only within the framework of a project, and they do not end when a project is completed. A project is limited in scope and time, community development is not. Hence, a flexible approach to planning is required. Frequently what is presented as a burning issue for local people may be well outside the scope of the project, as originally conceived in the project proposal. In Pascuales, Ecuador, for example, the brickmakers were keen to register their association in order to obtain land tenure and thus access bank loans (see Box 8.9). This was beyond the brief of the project, but unless they were given some assistance with this matter the brickmakers were unlikely to focus on what was the main project intervention.

It is probably impossible to forecast the ultimate consequences of any technical intervention. The process itself can generate new ideas and different priorities as capacity is increased. What appropriate technology organizations can do is to use accumulated knowledge and experience to avoid the more obvious pitfalls and to strengthen technical skills. This should leave communities in the position of owning the technology and being able to manage its on-going development.

PTD consists of a series of participatory activities, with related methods, which together comprise the key elements of technological innovation. Innovation takes place through focused and creative interaction between local communities and outside supporters. Continuous innovation is the core 'business' and requires continual interaction between the community and development professionals (Guijt and Veldhuizen, 1998).

The optimal degree of participation is specific to the context and the capabilities of the various stakeholders. It is also related to the stages in the process of technology development. The following sections show in more detail how PTD can be used as a project tool; how stages can be identified; and how progress can be monitored and evaluated (Lowe, 2001).

Degrees of participation

In terms of defining the degree of participation in technology development, a key question is: who has control? The balance of power in a project is critical. The relationship between the project promoters – outsiders – and the project beneficiaries – insiders – is fundamental. Outsiders must offer support but not foster dependency. The question is also one of ownership. Are people participating in the project belonging to outsiders? Or are outsiders supporting the people's initiative? These questions are not as simple as they may first appear. In order to begin to answer them, some analytical tools are useful. Categorizing the degree of stakeholders' participation is one way of gaining a better understanding of the complex realities (Pretty, 1974):

1 *Passive participation.* People are told what will happen, and information is only shared among outsiders.
2 *Participation in information gathering.* People are questioned, but the results of investigations and studies are not shared.
3 *Participation by consultation.* People are consulted, but outsiders define problems and solutions. However, these solutions may be changed in the light of people's reactions.
4 *Participation for material incentives.* People are co-opted into supplying resources or co-operating in return for material incentives.

5 *Functional participation.* People participate collectively in activities determined by external instigation and facilitation.
6 *Interactive participation.* People collaborate in analysis and planning. They take control of the decision-making process.
7 *Self-mobilization.* People initiate the project without external intervention. The project remains in their control.

There is no 'correct' degree of participation. Participation is specific to the context and the capabilities of the various stakeholders. It may also vary over the life of the project, evolving with the stages of technology development. The categorization of degrees of participation is in fact most useful when compared to these stages. It can give an indication of how the project is changing: is participation increasing or are outsiders increasing their control? This kind of analysis is a key element in an on-going system of monitoring and evaluation.

Even though there may not be an ideal level of participation, Box 4.2 illustrates the sort of pitfalls that arise when 'outsiders' try to impose their technological solutions without any participation by local brickmakers, other than 'land and labour'. This story was used in a workshop early in the brickmaking technology transfer project, which included ITDG staff from Peru, Zimbabwe and Britain, institutional allies and brickmakers from both Peru and Ecuador. The cautionary tale served its purpose. It reminded everyone – outsiders and insiders alike – of the pitfalls of technological arrogance, imported agendas and imposed 'solutions'. And it made the workshop participants laugh together!

> **Box 4.2** A brickmaker's experience of partnership
>
> My name is Adam. I'm a brickmaker, and the son of a brickmaker. In the dry season our family makes bricks and in the rainy season we farm. Four of us make bricks when someone – one of our neighbours or a person we know – needs them to build a house. Someone always needs a few bricks. A few years ago some guys came to our place. These guys asked a lot of questions. They took 'samples' of our soil in plastic bags and wrote labels on them. They took a few bricks away with them. They asked who we sold our bricks to, and at what price. Then they went away. A few months later they returned.
>
> 'Your soil is very good for brickmaking', they told us. They were right about that because we've been using this soil for many years. 'Your problem is you don't prepare your soil properly: you see the stones in this brick? Another thing, you're not burning your bricks correctly, not using enough energy. Look, they're soft, only dried mud really. The problem is you've used nearly all the fuelwood from around here, haven't you?' We have. 'And now you have to go a long way to gather wood?' We do. 'Deforestation and environmental degradation: big problems. You should use coal. Then you could produce very good bricks, and sell to the big building companies in town at a high price.' We could?
>
> These guys said they were from a development organization, an NGO. They had a big car, the best 4 × 4 by far, so they must be rich. They had something called 'donor money' from the European Community. We became partners, us and the NGO: partners in development. It was like this, we would provide the land and the labour and they would provide the money to 'upgrade' our brickworks.

They brought in four brickmaking machines all the way from Belgium. They put up a big shed so we could work through the rainy season. A soil-crushing machine arrived, also from Belgium. There was a mixer too. The machines had motors and needed petrol. 'No problem', they said, 'let's get to work'.

'But first, you need 42 staff. And at least 25 per cent must be women: it's in our Mission Statement'. Forty-two people? 'Look, we've worked it out on this chart: 10 digging, 10 preparing the soil, a team of 12 for the pressing machines, 7 building brick clamps and 1 site manager. That's you!' Oh. 'The work-force all need overalls, working boots, dust masks and hard hats. We mustn't be seen to be ignoring health and safety, eh?'

The process of soil preparation and moulding bricks with these Belgian machines was totally different to what we were used to. But they trained us; they told us what to do. The soil had to be crushed and sieved, then mixed with water. 'This type of machine needs the mix to be semi-dry,' they said, 'you can't press water.' They helped us build a huge kiln with six chimneys, big enough for 30 000 bricks! Normally we fire about a third of that. They had a book with a drawing of the new kiln inside. Just follow the plans, they said. They dropped in from time to time to make sure we were doing okay. And to have 'participatory project planning meetings', where they gave us our instructions for the next month.

This kiln was fired with coal. It was very new to us. They said it would make very good bricks, that the heat would be distributed evenly: it would be a model that all small-scale brickmakers in this country could adopt. The kiln cost a lot of money. 'But look, it's all here on this spreadsheet, it will pay for itself within 2 years.' It took us a long time to get the kiln working properly. In the end we had to block off five of the chimneys. Otherwise it belched smoke and flames like a dragon and all the bricks melted into one solid lump. We lost a lot of bricks.

'Teething troubles,' they said, 'research.' After that, things went okay for a while. They trained us to keep the records they wanted. They brought many visitors to see how well we were doing. They brought architects to marvel at the quality of our bricks. They brought a brickmaker from the other side of the world to look at our set-up, and brickmakers from other regions of the country to train alongside our workforce. They brought a government minister. He was *very* interested. 'Obviously money to be made,' he said, 'I might start a brickworks of my own.'

And then they left. 'Time to stand on your own two feet,' they said, 'you can't rely on donor money forever. Besides, we're off to the country next door: it's newly independent and democratic – flavour of the month with the donors.' But what about our partnership? 'Look, you know what to do. Order your coal, keep your records and good accounts, maintain quality control, pay attention to your marketing strategy – maybe look at diversifying your product base.' *Sorry?* 'Make tiles.'

Oh. But what about spares for the machines? What about transport – you used to bring in the petrol for the machines in your car? And what if the customers don't come? We don't have a telephone: you used to call them. 'Look, you're sitting on a goldmine. What you have here is a business with a potential turnover of a quarter of a million dollars a year.' But what if the customers don't buy? 'Bye!' And they left.

> There are four of us now. In the dry season we make bricks, and in the rainy season we farm. The kiln has cracked and fallen down, but at least with the shed we have somewhere dry to sit and talk about the days when we had so many visitors. Last week, some people in a four-wheel-drive vehicle came by. They asked if there were any brickmakers in the area. We said no.

Stages in technology development

It can be useful to consider six stages in the process of technology development (Haverkort et al., 1991).

1   *Getting started*. This phase should engage people in a relationship and establish a common understanding of roles and responsibilities. It is part of the process that is often overlooked and under-resourced. Technology development is a slow, step-by-step process, and it cannot be rushed. Short-term project horizons and the pressure to produce quantifiable outputs tend to crowd out the essential planning and initiating phases – for which projects inevitably pay later. At the start of the brickmaking project in Ecuador, for example, Alfredo Barriga describes how the purpose of the project was presented to the brickmakers of Pascuales, simply and clearly, and then the project team withdrew for a week to allow the brickmakers time to come up with their own reactions (see Chapter 8).

The initial interaction between insiders and outsiders is very important. It establishes how people perceive each other and what they expect from each other. There is a need to create an understanding of what each stakeholder brings to the project as quickly as possible. Sometimes it is advantageous to formalize this understanding by way of a memorandum of agreement or similar written document. Any agreement – written or verbal – needs periodic review if it is to retain currency.

2   *Looking for things to try*. Technology transfer projects spend a lot of time trying to understand problems and creating awareness of potential solutions – the options available to people to meet their needs. During this stage, collaborative research into these problems and solutions can be a powerful means of drawing stakeholders together, establishing a rapport and a common purpose. Development professionals are frequently from a different stratum of society to the people they work with; insiders and outsiders can feel very alienated from each other and even suspicious of each other's motives. The process of working together to understand existing technologies and the options available can help forge a mutually respectful relationship.

3   *Designing experiments*. It is often necessary and desirable for stakeholders to be convinced about the suitability or effectiveness of an improved technology. For example, it might be necessary to demonstrate to producers, local authorities and customers that the improved bricks being produced are strong enough to meet their requirements and standards. All these stakeholders need to have confidence in the product. Designing tests, particularly on-site tests, to ensure quality control is a process whereby the producer and adviser can co-operate fully. They will then see for themselves the benefits of all their hard work.

4   *Trying things out*. People must believe in alternative technological options if they are going to put in the effort to develop them. Where projects have incorporated letting people see for

themselves the merits of a technology, there has been a substantially increased adoption of that technology. The need for learning, however, includes an element of trial and error. The first coal-fired brick clamp that ITDG fired in Zimbabwe ended up as a solid cube of glass. Any risk has a cost associated with it. Sharing the cost of experimentation is crucial. Poor people and small entrepreneurs cannot afford to bear it alone.

5 *Sharing results with others.* Collaborative working, discussion groups, workshops, peer group meetings and exchange visits can all be effective mechanisms for communicating and disseminating results. They can involve and influence audiences beyond the immediate confines of the project. International dissemination, if appropriate, can be achieved via conferences and seminars, the use of print and electronic media. On a cautionary note, dissemination activities can involve significant costs and a lot of time. Furthermore, the information disseminated has to be appropriate to the audience. Information that is poor quality or is poorly presented will undermine and discredit the project. Communication is a specialized area and there is a case for employing professionals to undertake activities such as the written dissemination of information and video production.

On the other hand, there is an argument against constantly interpreting people's needs for them or putting words in their mouths: it may not, in many cases, be the most effective or equitable way of going about communication. Professionals may focus on demonstrating their cinematographic and editing skills to promote their own status, not capturing what people are saying – even when it may look bad. With this in mind, ITDG is involved with video projects that enable women to articulate their information needs in their own terms.

6 *Sustaining the process.* Project promoters, the outsiders, will almost invariably require a clear exit strategy. This could include helping to form producer associations or similar forums. It may involve developing the capacity of training institutions. The formation of linkages between stakeholders, independent of the project promoter, is important. The key is not to foster dependency and to help create an enabling environment for the continuation of the technology development process.

The role of external change agents – the 'outsiders'

External change agents may be appropriate technology (AT) organizations, other NGOs, technical specialists, government departments or local authorities. Whoever they are, they have to recognize the basic PTD goals of empowerment and equity. The following list covers a range of the roles that external change agents may accept. Not all roles will be relevant to every project, and the list is not intended to be exhaustive.

The facilitator:

- offers support in the strengthening of diagnostic skills, identifying technical and non-technical problems;
- helps with an analysis of how desired change may be facilitated by technology; and
- supports the community in tackling issues of power and vested interests where these are factors that affect the changes that are required.

The networker:

- strengthens the capacity to obtain information, both from inside and outside the locality;

- shares appropriate external information to enhance technical capacity; translates the technical needs defined by local people into specifications for suppliers and manufacturers; and
- assists in the formation of links between those wanting change and the market, R&D institutions, and other sources of knowledge and influence.

Table 4.1 An overview of participatory technology development with small-scale brickmakers in Peru

Stakeholder groups	Diagnostic stage: getting started, establishing common understanding, looking for things to try	Exploration of change options: designing experiments, testing options	Transfer of knowledge: sharing with others, sustaining the PTD process
Small-scale brick producers (in Ayacucho, Cajamarca, La Huaca and subsequently in Ecuador and Colombia)	Prior to interventions, producers reliant on fuelwood as primary source of energy. Early exchange with producers.	ITDG staff and producers in three locations; experimental kilns constructed, producers engaged in participatory production and firing trials; exchange visits between peer groups facilitate knowledge and skills development.	Peer group exchanges key to disseminating information about alternatives and practical skill sharing with producers in new locations. For example, Ecuadorian brickmakers visiting northern Peru to watch new technology hardware being used by brickmakers in La Huaca.
Academics (individuals and institutions focused on energy and SME issues)	International and national workshops facilitated exchange of information on key issues, constraints, and potential options.	Academics worked alongside brickmakers to explore issues of energy efficiency, measure production parameters and devise methodologies for monitoring production processes and improve product quality.	Dissemination via national and international workshops, publications and personal contact have shared research findings and triggered new initiatives in other locations.

(continued on next page)

Stakeholder groups	Diagnostic stage: getting started, establishing common understanding, looking for things to try	Exploration of change options: designing experiments, testing options	Transfer of knowledge: sharing with others, sustaining the PTD process
Government institutions (national and regional institutions in the construction and energy sectors)	Project engagement with individuals employed in government departments nationally enabled awareness to be raised through early dialogue.	Works with project, provides demonstration site and training facilities for male artisans.	Project research included in national statistics and raised awareness of the energy issues faced by small-scale producers has led to project partners being asked to brief ministers.
Artisans (supplying equipment to the small-scale brickmakers)	Face-to-face exploration of the issues affecting brick producers, as well as the potential for developing new technologies, has been effective in ensuring technological alternatives are rooted in local realities.	Small-scale manufacturers have been key partners in exploring practical options for the development of alternative hardware.	Small-scale entrepreneurs continue to spread the learning which the project facilitated. For example, one independent engineer working in the manufacture of oil-fired burners continues to support brickmakers by exploring solutions to the problems they encounter in employing new technologies.
ITDG project staff	Technical and social science staff from Zimbabwe, UK and Peru working with small-scale producers and other partners in seminars, practical workshops and sharing documented experience from other locations to increase exposure and build on existing experiential learning.	Project facilitates construction of numerous demonstration structures; some prove unpopular, e.g. rammed earth, others taken up independently of project, e.g. ferro-cement roofing.	Staff facilitate discussion workshops involving numerous stakeholders to complement own technology focus, e.g. health advisers, SME specialists. On-the-job training incorporated in construction of demonstration houses.

The educator:

- communicates knowledge for the identification of changes and solutions;
- helps extend the range of available solutions; and
- offers advice when requested, but is prepared to withdraw when not needed.

The co-researcher:

- helps identify the non-technical skills that are needed; and
- recognizes and helps to develop local technical skills in experimentation and adaptation, underwriting risk if necessary.

Building technological capability

Technological capability can be defined as the ability to select appropriate technologies, to absorb and adapt technologies into local settings, and to develop new processes and products through local innovation. ITDG believes that technology development requires the recognition of local knowledge and skills – industrialized countries cannot impose it. The following are considered key factors in building technological capability (Biggs, 1995).

Formal education and training. Generally, training courses should combine theoretical and practical sessions. They have to be tailored to the educational level of the participants. It should be noted that education in entrepreneurial skills – business management, marketing, bookkeeping and so on – is frequently required as a follow-up to technological training. These skills are at least as important to the success of a productive enterprise.

Learning on the job – learning by doing. Many people prefer an experiential 'hands-on' approach to learning. This is not surprising, given that the people that appropriate technology organizations work with often have not had much formal education. Their work is practical and so they are more comfortable with a practical approach to learning. On-the-job training, whereby project promoters work side by side with artisans, can be a mutually educational experience (there's nothing like getting a middle-class engineer covered in mud to make him or her fully appreciate the rigours of the brickmaking process). In order to involve women, projects must recognize the constraints of their domestic, community and productive roles. Training schedules should be tailored to fit in with women's other commitments.

Integration of past experience. Integrating existing knowledge is critical to the process of technology adoption and adaptation. Facilitating the exchange of knowledge between producers in the same sector can stimulate significant advances.

Entrepreneurship. Technology transfer work usually takes place with the owners of small- to medium-scale enterprises (SMEs), who already have experience in the manufacturing process and in running a business. While there are distinct advantages to working with functioning and defined businesses, these people are not usually amongst the 'poorest of the poor' in their own countries. The justification for working with the 'not so poor' is that the wider social and economic returns derived from the project will exceed the benefits to the employers themselves. Jobs will be created and the additional income generated can be distributed equitably. On the national level, the development of indigenous SMEs reduces the need for imports, thus saving on scarce foreign exchange. Working with SMEs can lead to a more dynamic, diverse and competitive private sector.

For SMEs, technological capability is defined in terms of the ability to recognize market opportunities and to source the raw materials, skills and equipment to exploit them. Since a lack of technological capability is frequently the most important constraint on SMEs, the provision of business development and support services is obviously very important.

Institutional support. If SMEs need business support, then often business development service institutions need support too. Training institutions, producer associations, business advisory bureaux and service centres are all liable to require assistance to provide the services necessary to sustain technology development. Public sector agencies and government departments in developing countries commonly need support; they are frequently underfunded and understaffed. Project promoters often search out the services that producers need, only to find these services themselves need support.

Cultural environment. Different approaches and tools must be employed in order to work effectively with the diversity of social, cultural, educational and individual differences that every community encompasses.

Regulatory environment. In order to allow poor people access to affordable technologies and an outlet for their products, the public sector has to provide an enabling environment. Project promoters may well find that they have to play a lobbying and advocacy role on behalf of small-scale producers. Outdated legislation, leaden bureaucracy and corruption can prove major stumbling blocks to technology and business development. Assisting the formation of producer associations – or similar organizations – to undertake this lobbying role should be considered as a long-term strategic goal.

Conclusion

The parameters of the process of PTD have been defined. The degrees of participation in the process have been suggested; the stages of technology development have been outlined; and some key factors in building technological capability enumerated. Categorization of the elements in action can serve as a useful tool in analysing a project and its impact. At any stage of technology development, for example, the project team can attempt to ascertain the degree of participation of the beneficiaries: they can construct a matrix to give a snapshot of the project and its progress at that point in time. Some projects may remain static in this regard, while others exhibit an increasing participation. As was stated initially, the optimal degree of participation is specific to the context and the capabilities of the various stakeholders.

At any stage in the technology development process, the team can look at how well the key factors in building technological capability are being addressed. Once again, it will depend on the nature of the project as to which factors are addressed when. They can also look at the factors in conjunction with the evolving degree of participation.

As with all investigative techniques, the picture that emerges will be dependent on a host of variables. It will be subjective; it will depend on who is asking the questions, who is answering them, what is asked and how it is asked. Nevertheless, the end product – the snapshot matrix – should give a useful overview (see Table 4.1). It is the process of producing the matrix, however, which may be most important: it will inevitably stimulate debate and may save the project from veering irredeemably off course. In other words, it has the potential to add value for the stakeholders within the PTD process. In project planning, monitoring and evaluation, using these PTD tools can be very valuable indeed.

With increased emphasis being placed on communities defining and controlling the developmental process, there is a tendency to undervalue the role of external change agents – the outsiders. The implicit suggestion is that communities have all the necessary resources to help themselves out of poverty; all they need is mobilization or facilitation, there is no need to import 'experts'. While there is clearly no single optimal solution, each situation can benefit from both internal and external human resources. If these are brought together in a collaborative process that ensures poor people maintain or increase their control over their lives, then PTD can work both ways to enhance mutual understanding and empowerment.

PART II

THE BRICKMAKING TECHNOLOGY TRANSFER PROJECT

5
THE CHOICE OF PERU AND ECUADOR

AT THIS POINT, the brickmaking technology transfer project in Latin America is introduced. The justification for intervention is discussed, and the principal objective of the project stated. The relationship between poverty and the environment is examined. ITDG's work with brickmakers around the world is summarized and some of the Group's experience with South–South technology transfer is presented. Thereafter, following a general introduction to the countries, the situation vis-à-vis housing and brick production in Peru and Ecuador is examined.

Why Peru?

In 1996, the ITDG began working with small-scale brickmakers in Peru. The decision to start this project was preceded by studies of the needs of brickmakers (Mason, 1994), the national demand for building materials and the state of the brickmaking industry in general. Three key points emerged which persuaded ITDG that its intervention would be valid.

First, it was clear that the wood-fired Scotch kilns used by most small-scale brickmakers in Peru were comparatively inefficient. Throughout the country, brickmakers were using too much fuel to fire their bricks. Furthermore, the cost of fuel was typically half the cost of production, so this inefficiency significantly affected the income of brickmakers. Small-scale brickmakers generally supply the bottom-end of the market, customers whose main criterion is low price rather than high quality. Thus, brickmakers operate on a very limited profit margin. In most areas it is a buyer's market, and brickmakers cannot simply increase their prices in order to increase their income. The cost of energy inefficiency is, therefore, critically important to the viability of small-scale brickmaking enterprises.

The second key point was that fuelwood was in very short supply, and in many areas it could only be obtained illegally. There is a national problem of deforestation, subsequent soil erosion and environmental degradation. Hence, legislation is in force to protect indigenous woodland, and there are heavy fines for anyone caught stealing wood. Particularly in the most vulnerable areas, the police and forestry officials are active in enforcing the law. These areas are inevitably the places where many brickmakers operate: deforestation is most serious where there is a high population density, and where there are a lot of people there is a high demand for housing, and hence bricks. Brickmakers are aware of the environmental consequences of cutting fuelwood, and of the laws prohibiting it. They are acutely aware of the penalties – fines that they cannot afford to incur. They have no choice, however, but to take the risk and further damage their own environment. Like many poor people in the Third World, they are the victims of environmental degradation as well as its perpetrators, and this is because of an absence of alternatives (see Box 5.1). Sufficient fuelwood from other sources – managed woodlots or waste from sawmills – is generally either unavailable or unaffordable.

Finally, the situation for small-scale brickmakers throughout Latin America was known to be similar to that in Peru – though of course there were national and regional differences in technology. Fuelwood shortages, deforestation and energy inefficiency are, in fact, common in much of the developing world. A comparative study of communities in India and Ghana, for example, found that, unsurprisingly, fuelwood scarcity was a key energy constraint in both areas; it affected all members of the community, not least brickmakers (Wilkinson, 1999). One of the aims of the project in Peru was to develop a clearer picture of the common problems affecting the small-scale brickmaking sector in Latin America, in the hope that some of the more fuel-efficient methods of brick firing developed in Peru would be applicable elsewhere in Latin America.

To extend the validity of the project, ITDG undertook parallel and complementary work in Ecuador. The situation in Ecuador was somewhat different to Peru: apart from differences in culture and socio-economic conditions, brickmakers used different brick production and firing technologies (Barriga et al., 1992). In order to expand the project's relevance further, a number of international workshops were proposed to bring together those working at the cutting edge of brickmaking and energy efficiency, especially stakeholders from Latin America.

A possible entry point: the coal-fired clamp

ITDG had experience of brickmaking, using a variety of technologies, in several countries in the developing world. In Zimbabwe the principal technological innovation had been a coal-fired clamp (Mason, 1993). This was reportedly much more efficient than the Scotch kilns commonly used in Peru, so it appeared to be a technology that would suit the needs of the brickmakers. At the very least, it could be an entry point: a proven technology that was not radically different to current practice and that did not require a major capital investment. Adapted for Peru, the coal-fired clamp could increase the efficiency of the brick-firing process, thus cutting production costs and increasing brickmakers' incomes. In addition, it would use an alternative fuel that preliminary research indicated was available to many. Whether or not the technology could be successfully transferred to Peru was the main question that the project set out to investigate.

The coal-fired clamp developed in Zimbabwe was an adaptation of a well-known but previously industrial-scale western technology. Earlier work with the coal-fired clamp in Zimbabwe should yield valuable lessons for the start of the project in Peru (see Chapter 6).

> **Box 5.1 Poverty and the environment**
>
> From the vantage point of the poor, there are good reasons why poverty exacerbates environmental damage. Increasing impoverishment and the absence of alternative livelihoods mean that an ever-growing number of poor people exert unprecedented pressure on the natural resource base as they struggle to survive under conditions of risk and uncertainty. Poor people and environmental damage are often caught in a downward spiral. Much of the environmental damage that affects the poor is beyond their control. Past resources degradation deepens today's poverty. Today's poverty, however, makes it hard to find sustainable economic alternatives to deforestation, which would prevent further desertification and loss of biodiversity, and to control erosion and replace soil nutrients in fertility-depleted arable crop fields.
>
> It is axiomatic, therefore, that those who live in poverty are forced by circumstances to deplete resources in order to survive. It is unfortunate, however, that this degradation further impoverishes these people.
>
> *Source:* Chigaru (1999)

The situation in Zimbabwe was markedly different, however; not only did Peruvian brickmakers not use the same technologies, but their culture, the market and the way brickmakers organized themselves all differed between Zimbabwe and Peru (see Chapter 7). In short, many factors would critically influence the take-up of the technology. It was evident from the outset that the project would not merely be a matter of presenting brickmakers with a predetermined or 'stock' technical solution.

The project objective

From ITDG's point of view, the interest in the project in Latin America was not principally in researching and developing technologies for small-scale brickmakers. It was thought that a good deal of that work had already been done. *Thus, the main purpose of the project was to develop and disseminate a technology transfer methodology in energy-efficient techniques and technologies appropriate to small-scale brick production.*

ITDG wished to research the institutional framework and the dissemination processes necessary to spread knowledge on a wide scale. The methodology would be developed from the parallel research programmes in Peru and Ecuador.

ITDG already had considerable experience of South–South transfer of technology (see Box 5.2). Building on this experience (e.g. in the food-processing sector), the challenge with brickmaking technologies was to develop a methodology for the technology transfer and development process. It was intended that the methodology would be relevant to technological areas apart from small-scale brickmaking and to sectors of the economy other than the building materials production.

Following discussions with stakeholders in Peru and Ecuador – the brickmakers, potential institutional allies, policy and decision makers – it was decided that there was no conflict of interest between the brickmakers' needs and ITDG's wish to evolve a technology transfer methodology. The 'agendas' meshed well: the technology transfer methodology would be driven by the brickmakers' needs. Because the project wished to develop a

> **Box 5.2** South–South technology transfer: the Tinytech oil mill and the tray dryer
>
> ITDG had previous experience of transferring technologies between countries in 'the south', notably in the area of agricultural and food processing. One example is the Tinytech, an Indian oil mill, which was introduced to Zimbabwe. The mill is used to produce cooking oil, principally, in the Zimbabwean instance, from sunflower seeds. The dissemination strategy was via local entrepreneurs whose attention was attracted by the publication of an investment prospectus in national newspapers. ITDG then assisted the entrepreneurs with the installation, commissioning and running of the mills. The Tinytech worked well in the pilot phase and three entrepreneurs were engaged in oil production, employing a number of people and supplying a burgeoning market. The project then investigated the potential for manufacturing the mill locally.
>
> A second example is tray-drying technology. In the early 1980s, ITDG's agro-processing programme identified the potential for a small-scale, low-cost tray dryer that could be used to dry herbs, teas, fruits, vegetables and flowers. Dryers were available in Peru, but they were designed for large-scale industrial use. These dryers had a sizeable product capacity, required huge volumes of air to be heated and were too great a capital investment for them to be suitable for small-scale operations. ITDG identified a tray dryer design that was being successfully used in Guatemala. Alliances were formed with a local engineering company keen to extend its product range, and an entrepreneur interested in starting a drying business. The entrepreneur paid for the construction of a dryer, while ITDG provided the imported heater-blower on condition that the technology could be monitored.
>
> The technology worked well. The engineering company became a technology advocate, advertising the dryer in the press. Between 1984 and 1996, 65 dryers were sold in Peru, mainly to small- and medium-scale food processing firms who supplied the urban market. Each tray-drying installation typically created two jobs in its direct operation. In addition, post-drying processing could create up to 10 jobs, usually for women.
>
> Accounts of technology transfer tend to focus on the hardware. It should be noted, however, that the software – education, training, technical and business support – are vital elements in the process.

methodology that would be widely applicable, contact was also made with people working with brickmakers in other countries. Alliances were formed for information exchange and co-operation. (Latterly, some practical extension work was also undertaken in Colombia.) Knowledge of brickmaking in other Latin American countries, and indeed globally, would ensure that the project's findings were relevant not only to Peru and Ecuador.

The situation in Peru

Peru is a mountainous republic on the west coast of South America. It occupies an area of around 1 285 000 km^2. The population is in excess of 25 million people. Annual population growth is around 2 per cent (World Development Report 2000/01); life expectancy is approximately 63 years for women and 60 years for men. The capital city, Lima, is home to more than 25 per cent of the population, about 6.25 million people. This number is increasing rapidly and the city sprawls over a large area, with many outlying informal settlements or shanty towns.

A majority of the population is indigenous Indian of the Quechua and Aymara tribal groups (around 45 per cent). Europeans and mixed-ancestry *mestizos* account for the bulk of the remainder, around 37 per cent. Most Indians lead traditional lifestyles, living in quite remote Andean villages and making a living through subsistence activities. This relative isolation has, to a large extent, kept them outside the mainstream of Peruvian national and political life. The Quechua and Aymara have their own languages and many Indian villagers speak little or no Spanish. The language of politics and commerce, however, is Spanish.

Peru has abundant mineral resources and is one of the world's leading producers of silver, zinc, lead, copper, gold and iron ore. Around 80 per cent of the country's oil is extracted from the Amazon forest. Government in Peru has been notoriously unstable, but a democratic constitution has been in place since 1980 and President Alberto Fujimori had been in power since 1990. The most recent period of violent political unrest ended in 1993 with the quashing of the Maoist guerrilla group, Sendero Luminoso or 'Shining Path'. The country is predominantly Roman Catholic, and literacy is around 85 per cent.

An estimated 56 per cent of the population, including most of the rich and middle class, is concentrated on around 10 per cent of the land area in the narrow coastal region, which has reasonably mild temperatures all year round. It comprises arid plains and foothills with areas of desert and fertile river valleys. Most industry and commerce is practised in this region, with a particular concentration around Lima. Cash crop production is also concentrated along the coastal fringe. Particularly important for the economy are coffee, cotton and sugarcane.

The *sierra* or mountain region, which has an average altitude of 3000 m, is home to around 36 per cent of the population. The *montaña selva*, or jungle region, consists of the forested Andes and Amazon basin. This region is home to the remainder of the population. It has a typically wet, tropical climate.

Housing and brick production in Peru

There is a large-scale housing problem in Peru. In and around urban centres the situation is critical, with large swathes of poorly housed people, inadequate services and infrastructure. Formal urban construction utilizes typical modern materials, primarily bricks and cement. Informal settlement dwellers build with whatever is freely or cheaply available: tin, plastic sheets, waste timber and cardboard.

In the rural areas of the coastal region, most houses are built of traditional adobe: sun-dried clay. Other houses are constructed with locally produced fired-clay bricks. A combination of adobe and stone is the principal construction technology in rural areas of the mountain region. In jungle regions, houses are mainly built from wood. *Quincha*, a type of wattle and daub, has been used in parts of Peru for centuries. Traditional *quincha* houses exhibit an inherent earthquake resistance. As Peru is prone to earthquakes, it is a good choice of technology in many situations. Following a devastating earthquake in 1990, ITDG has pioneered the development of improved *quincha* in alliance with local communities (Lowe, 1997).

Lima is the main centre of the industrial-scale production of fired-clay bricks. Small-scale production of bricks is, however, also important in Lima, as in all regions of the country.

These 'artisanal' bricks are used mainly in the construction of private houses for the less well off majority of people. Close to the cities and large towns, bricks made on an industrial scale are cheaper; however, small-scale production is most important and most viable in areas where the cost of transport makes industrially produced bricks unaffordable or unavailable. The small-scale brickmaking sector uses an estimated 80 000 tonnes of petrol equivalent (TPE) per annum, around 0.65 per cent of Peru's total energy consumption (although accurate estimates are notoriously difficult to come by for the informal sector).

It is estimated that there are more than 2500 small-scale brickmaking enterprises in Peru, employing, on a full- or part-time basis, more than 10 000 people. (This is ITDG's estimate; no other estimates are available, and in fact ITDG Peru provides the Ministry of Energy with the information from which it compiles statistics on the sector.)

The project concentrated on four regions where small-scale brickmaking was particularly important: around the town of Ayacucho, which is in the mountain region, south-east of Lima; in the La Huaca area, near the northern town of Piura, where brickmakers operate in an arid region of the coastal belt; near the town of Cajamarca is in the mountain region, north of Lima; and the region of Alto Mayo, on the fringe of the jungle region, near the town of Moyobamaba, which lies north of Lima.

The situation in Ecuador

The republic of Ecuador, which includes the Galapagos Islands with their unique species of flora and fauna, has an estimated population of more than 12 million people. Less than 10 per cent of the population live in the jungle provinces and the Galapagos; and the remainder is divided roughly equally between the lowlands and the highlands. Approximately 25 per cent of the people are Indian, 55 per cent *mestizo*, 10 per cent Spanish and 10 per cent African. Life expectancy is around 68 years for women and 63 years for men. The population is growing at a similar rate to that in Peru. Agriculture employs almost 50 per cent of the workforce. The most important crops – exported to earn vital foreign currency – are coffee, bananas and cocoa. The main religion is Roman Catholicism; the principal and official language Spanish; and literacy is similar to Peru at 86 per cent.

Set on the north-west coast of South America, the country has a land area of around 270 000 km^2. The terrain varies from coastal plain in the west, descending from rolling hills in the north, to a broad, lowland basin. In the centre, the Andean uplands rise to snow-capped peaks. In the east, forested alluvial plains are dissected by rivers, which flow from the Andes towards the Amazon. The equatorial climate is wet and warm all year round, though obviously cooler in the mountains. The capital city, Quito, is home to around 2.3 million people. The other major city, Guayaquil, which is the principal mercantile centre, has an even larger population of almost 3 million. There are ten cities with between 200 000 and 500 000 people; and the total urban population lies between 7.5 and 8 million.

Economically, Ecuador has fallen on hard times since a brief boom in the 1970s that was fuelled by the country's oil reserves. Low oil prices and a large foreign debt mean that Ecuador struggles for economic stability. This is reflected in the political instability that has plagued the country since it became a republic in 1830, gaining independence from Colombia. Prior to 1992, Ecuador had seen 15 changes of government in 40 years. In February 2000, an uprising among the Indian population, the poorest sector, resulted in the abdication of the president and his replacement by the vice-president. There has been a long-running dispute with Peru over border territories in the rainforest. The discovery of oil in this region added to the tension and there was a war between the two countries in 1981. The present border is now grudgingly accepted.

Ecuador exports raw materials and products. Total gross national product is around US$16 000 million, though because of the on-going economic crisis this is at best a rough estimate. The principal exports are: petroleum products (worth US$1300 million); shrimps (US$800 million); bananas (US$600 million); and cocoa and coffee (US$500 million).

Housing and brick production in Ecuador

According to government reports, the housing deficit runs at between 500 000 and 1 000 000 units. It is difficult to obtain a more accurate estimate, and some observers feel that even the higher figure is an underestimate. The situation is complicated by heavy in-migration from rural to urban areas.

A wide variety of construction materials and techniques are used in urban areas. There are numerous high-rise buildings built with modern construction technology, employing structural steel. A city like Guayaquil may have 20–25 per cent of family housing built with full masonry and modern finishes. Another 40–45 per cent, the housing of the working class, is built with a mix of cement blocks and bricks. The remainder includes low-cost housing made with artisanal bricks or bamboo. Urban slums surround Guayaquil and other cities, and in these informal settlements, bamboo is the principal material used to construct houses. After a few years, however, bricks and cement almost invariably replace bamboo.

There are two or three large industrial manufacturing plants that supply the market with bricks. Industrial fired-clay bricks are second only to commercially produced cement blocks as the walling materials of public choice. Artisanal clay bricks are also sold, but they are not well regarded due to their poor finish, variable shape and sometimes over-burnt appearance. They are used for non-load-bearing walls and in low-cost, low-rise housing. Even for low-cost government programmes, however, good-quality cement blocks are preferred because they need no external cement rendering. Artisanal brick walls have a less 'pleasant' look and need rendering to give an acceptable finish and prevent excessive water absorption when exposed to rain. The main artisanal brickmaking centres are located around the city of Guayaquil and in Chambo in the highland province of Chimborazo.

Deforestation is a serious problem in Ecuador. Reports indicate that the rate of deforestation between 1980 and 1995 was more than 150 000 ha per year. Oil exploitation activity in the jungle, the expansion of agricultural use and some wood extraction for domestic consumption were said to be the principal causes. Nowadays, wood chips are also exported to Japan, mainly for paper manufacture, and this is the cause of extensive tree felling. In Chimborazo, brickmakers complain of a shortage of wood owing to the demands of the chipping factory in northern Ecuador, which processes wood for export to Japan.

It is estimated that in Ecuador there are up to 2500 brickmaking enterprises, distributed throughout the country. Techniques of production differ from region to region. Throughout the country, a variety of species of fuelwood – depending on location and availability – are used as the principal fuel for brickmaking. If it is assumed that there are 2500 brickmaking enterprises using fuelwood in Ecuador, and that each has a firing approximately four times per month, and each firing uses at least 300 kg of wood, then the national monthly consumption can be estimated as 3000 tonnes (36 000 tonnes per year). This obviously has a major impact on forest resources.

6
EARLY EXPERIENCE IN ZIMBABWE – THE COAL-FIRED CLAMP

BEFORE PROCEEDING with the story of the project in Peru and Ecuador, it is interesting to look in some depth at ITDG's experience of working with brickmakers in Zimbabwe (this account is based on Mason, 1997). An evaluation of the work, which drew out the lessons learned (Jazdowska, 1997), was studied in detail by project staff in Peru before they began to design their project. The evaluation was carried out bearing in mind the issues of technology transfer, appropriate technology and PTD.

Brickmaking in Zimbabwe – the historical context

Prior to independence in 1980, Zimbabwe was a colonial country and its society was divided on racial lines. On the whole, how you built your house was determined by which racial group you belonged to. The commercial demand for fired-clay bricks was mainly from the economically dominant white minority, and a few large brickworks situated near the major cities met this demand. These plants are still in production, though in most cases the buildings and machinery are old and failing. The brickworks employ industrial-scale technology: mechanical clay mixing, extrusion moulding and permanent kilns.

Historically, the indigenous rural population did not use burnt clay bricks in the construction of their homes; walls were traditionally built using 'pole and dagga' (wattle and daub), whereby a wooden structure was plastered with mud. In colonial times, however, producing 'farm bricks' became a commonplace activity throughout the country. Farm bricks were used mainly on the commercial farms owned by whites, although it was the indigenous workforce who actually made the bricks. The technology is therefore currently well known and quite standard throughout the country; in fact, it has become known as a traditional technology.

Farm brickmaking involves 'slop moulding' clay: that is, digging the clay, mixing it with water and forming it while very wet using a simple wooden mould. Typically, the production site is chosen to be near a large anthill, which provides a source of clay, and close to a supply of water. Once moulded, the bricks are laid on the ground in the sun. When dry, they are stacked into a clamp of the 'scove' type – a truncated pyramid of bricks with firing tunnels built into it. The number of bricks in a clamp varies from a few thousand to perhaps 20 thousand. The clamp is scoved – plastered with mud – to provide thermal insulation. It is burned by lighting wood fires in the firing tunnels. These fires are maintained for a day or so before the kiln is allowed to cool. The quality of farm bricks is generally poor. They are

Photo 6.1 A brick clamp being fired in Rwanda.

Theo Schilderman/ITDG

misshapen, under-fired, relatively weak and absorb water too readily. Nevertheless, they are perfectly adequate for the construction of single-storey houses and non-load-bearing walls.

Post-independence, society changed significantly and so did the demand for bricks and other building materials. More people than ever began to move to urban centres in search of work. But the cities were overcrowded and many of these migrants were housed in 'satellite' towns and 'high-density suburbs' (townships). The typical high-density house is single storey, with perhaps four small rooms. Walls are built from fired-clay bricks or concrete blocks.

By the mid-1980s the housing shortage was acute. Residents of high-density suburbs commonly rented out rooms in their already crowded homes. Alternatively, they rented a part of their small plot for another family to construct a shanty house, or *tangwena*. Demand for bricks far outstripped supply, and consequently the price increased markedly. Similarly, cement for making blocks became prohibitively expensive. The shortage of walling materials meant that people turned to farm-brick producers. Zimbabwe had, however, inherited a legislative standard for brick quality from the British colonial administration, which effectively prohibited the use of farm bricks in urban construction. The quality of brick required for general building purposes in towns and cities was defined as 'common' and, in general, farm bricks did not meet this common standard for strength, water absorption or appearance (see Box 6.1).

For political and economic reasons, there was little investment in Zimbabwe during the 1980s. Neither government nor private investors had the capital to build large-scale brickworks to supply the demand. Some indigenous entrepreneurs – perhaps a handful nationwide – did start brickmaking enterprises on a medium to large scale, but very few were successful. The technology that these entrepreneurs chose was often highly mechanized

> **Box 6.1** Building standards for bricks in Zimbabwe
>
> Bricks categorized as 'common' by the Standards Association of Zimbabwe have a crushing strength of 7 MPa. Under specified test conditions they have a water absorption of less than 15 per cent by weight, and resist a spraying test for erodability. In addition, the shape is regular and the dimensions conform to those of the standard. Bricks that do not meet this standard are classified as 'farm'. Above the common standard are a variety of 'face' and 'industrial' bricks that have a crushing strength of 15 MPa and some form of attractive finish. Then there are engineering bricks, with a crushing strength of 35 MPa, and finally a range of 'specials' such as refractories and insulation bricks.

and imported, and this, together with a lack of technical knowledge and management experience, contributed to the failure of some of these businesses. Though others survive and prosper, they are so few that they have not had a significant impact on the national shortage of bricks. Meanwhile, a number of producers in urban-fringe and peri-urban areas started to make farm bricks to supply the urban market. In defiance of the regulations, these bricks were used in the construction of urban housing.

The choice of the coal-fired clamp

It became apparent that there could be a role for small- and medium-scale producers to supply the urban market with bricks of an acceptable quality. These producers did not have – and could not raise – money to invest in mechanization; at the same time, unemployment was very high and labour affordable. Considering this socio-economic situation, intermediate technologies were judged to have the potential for success.

Following extensive consultations with stakeholders, the chosen technology to be introduced was the coal-fired clamp. This clamp needs no permanent structure, and features the use of coal placed between the layers of bricks (see Figure 2.1). There were several reasons for choosing this technology. First, the coal-fired clamp was a proven technology, employed worldwide, in which bricks could be burned to the required standard. Second, compared to most batch production processes, the coal-fired clamp was efficient – much more efficient than the scove clamp. Third, having no permanent structure, the clamp required no major investment. There was also a pressing need to discourage the use of wood as a fuel. Zimbabwe was – and still is – experiencing major environmental problems associated with deforestation. Fortunately, the nation has its own supply of coal and it is readily available around most urban centres.

As work with brickmakers evolved, ITDG considered all aspects of the brickmaking process: site planning, business planning, site operation, marketing and the environmental impact. Experiments were undertaken with various technologies for clay crushing, soil preparation, brick moulding and so on. The coal-fired clamp, however, remained the pivotal technology. In a reasonably short time it was adopted by a number of brickmaking enterprises. These enterprises varied from training establishments to co-operatives and emergent entrepreneurs. The clamp proved a suitable choice for brickmakers over a wide range of scales of production. It was a flexible technology: the size of clamp can vary from, say, 15 thousand bricks up to an undefined maximum. In theory, the larger the clamp, the

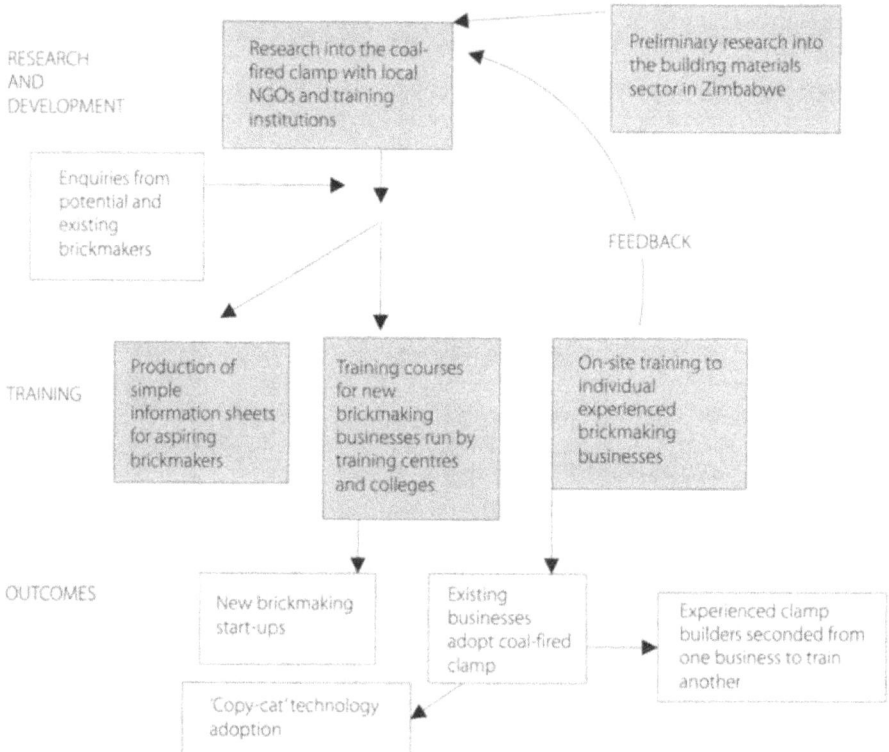

Figure 6.1 The technology transfer methodology developed in Zimbabwe

more energy efficient the firing process. Typically, brickmakers in Zimbabwe build coal-fired clamps of 20–30 thousand bricks.

The process of technology transfer

The dissemination of the coal-fired clamp in Zimbabwe can be regarded as the transfer of a well-known technology. However, the large-scale, coal-fired clamp common in Europe certainly had to be adapted to suit small-scale producers in Africa. The technology did not come with an instruction booklet. It had to be experimented with and adapted before project staff had the experience and confidence to begin dissemination. Work with local NGOs and training institutions provided the opportunity to gain this experience and confidence within a research environment. Hence early failures did not result in financial losses for small producers, nor did they discredit the technology among potential users.

Once dissemination of the technology began in earnest, ITDG adopted a diverse approach. While continuing to work with other NGOs and institutions, they began to respond to the many enquiries coming from brickmakers. ITDG maintained a planned and structured programme of work with its institutional partners: developing clamp firing technology; experimenting with the substitution of waste products and residues for coal; and training a variety of individuals and groups in clamp building and firing. One particularly successful innovation has been the substitution of boiler waste for coal. Boiler waste was available free from Harare's aged coal-fired power station. The station actually had a problem with disposal and brickmakers had only to pay for transportation. The boiler waste retains a high calorific value due to inefficient combustion in the power station's boilers.

In the case of aspiring brickmakers, assistance often had to begin at a fundamental level: testing clay and estimating the extent of reserves. In a short time the number of enquiries was such that ITDG produced a range of simple information sheets. These were photocopied and distributed free of charge. Following up on serious enquiries meant ITDG had to adopt a very reactive and flexible approach. The nature and source of the enquiries varied enormously: from businessmen asking about international sources of brick-moulding equipment for a major plant; to a small cooperative – who did not even have land – enquiring about the possibility of starting a brickmaking venture.

With respect to training, ITDG adopted a range of approaches. 'Formal' training courses were run with institutional partners at training centres and colleges. Meanwhile, ITDG staff provided intensive on-site training to individual brickmaking businesses. The strategy of seconding an experienced clamp builder from one enterprise to another was particularly successful. In short, a flexible and diverse approach evolved that was responsive to the needs of individual producers. While remaining proactive in research and development work, ITDG tried to react to the diverse and changing needs of enquirers. It was a difficult balancing act.

Successes and feedback

The ITDG project dealing with building materials production in Zimbabwe started out as a small one. When work with brickmaking started in 1987 there was only one staff member working on building materials – and even he was not full time. From 1990 to 1992 the project team grew to five people, four of whom were working almost exclusively on various aspects of building materials production. This brief appraisal of the project, the technology, and its transfer does not do justice to the size of the programme.

Level of adoption. In October 1996 it was estimated that more than 60 separate businesses were using the coal-fired clamp. There may well have been more: it was difficult to keep track because dissemination had become a 'copycat' process, with one producer learning from another, without the involvement of project staff. These businesses would not have adopted the coal-fired clamp unless it was cheaper or produced a better-quality brick that would command a higher price. This level of adoption – effectively achieved in just six years – is probably a good indicator of a successful technology and reasonably successful technology transfer.

Understanding the local building materials sector. Being well informed about the building materials sector was necessary for the success that the project achieved. Data gathered at the beginning of the project provided a good indicator of brickmakers' needs, and most of the assumptions made before work started were likely to remain valid for a long time to come. These included: that a market exists for improved bricks in urban areas; that a technology that saves energy – and money – would be adopted by brickmakers; and that small-scale brickmaking could create employment and generate income.

At the same time, it was also important that the project did not adopt a rigid approach because the needs of the brickmakers were not fully known at the outset. Finding out what these needs were was a process that took time and the experience of working closely with them. Therefore, the bulk of the work had to be dynamic and responsive: there was no rigid list of rules and stages to be followed day by day, instead the work plan had to remain flexible.

Geographical concentration. Unsurprisingly for a project that kept its staff very busy and that moved so fast, there have been areas where the feedback is not entirely positive. In general, the people who have taken up the clamp technology are distributed within a 150 km radius of Harare, the capital city. Zimbabwe is a large country with several diverse geographical regions where the coal-fired clamp was not adopted, and thus the technology cannot be said to have been proven nationally. At the time of the project there was only a small project team based in Harare, and this is why adoption of the coal-fired clamp was concentrated around this city. Only within the last few years has ITDG expanded to have offices in other regions.

Shortage of data monitoring energy usage. Rigorous and scientific monitoring of the energy use of the coal-fired clamp and its 'competitors' was not achieved. Data were collected in a haphazard way. Although the data *suggest* that the clamp is an energy-efficient and cost-effective technology, its performance could not be presented in a way that was directly comparable with other technologies. Time, staff shortages and the dynamic nature of the project did not allow for the methodical collection of data.

There was another very important global reason for this lack of comparable information. *There was no international standard for the collection and presentation of data to compare energy use in brick production.* In short, data collected in terms of, for example, x cubic metres of firewood used to burn y thousand bricks is virtually meaningless. Around the world the calorific value of firewood varies, as does the size and mass of bricks produced. This was a problem that the project in Latin America had to address at the outset by developing a methodology for monitoring energy consumption for a kiln (see Chapter 7).

Lessons learned

There are a number of elements that can be identified as having contributed to the successes of the project. Encouragingly, these compare well to the tenets of technology transfer, appropriate technology and participatory technology development.

The research and development of coal-fired clamp technology was carried out in Zimbabwe, drawing on international knowledge where necessary. This is an important point, because experience has been gained locally at all stages of development. Project staff had good, first-hand knowledge and experience. They have actually made bricks, fired clamps and seen the problems.

Initial experimentation should take place within a sheltered research environment. This ensures that neither the technology nor the project staff is discredited by any possible disasters during the experimentation phase. Differentiating between research work and the dissemination of proven technologies was therefore a key element of success.

Appropriate technology. The coal-fired clamp has undoubtedly proved to be a success among brickmakers in Zimbabwe. It has been quite widely adopted, and brickmakers are choosing it over other methods of firing for the following reasons:

- the bricks fired can be of good quality;
- the process seems to be energy efficient – thereby reducing the cost of production and the emissions associated with burning any fuel;
- fuelwood resources are being conserved by the more widespread use of coal or boiler waste;
- a relatively small change in technology was required in changing from the scove kiln to the coal-fired clamp;

- the coal-fired clamp is labour- rather than capital-intensive. In fact it has no capital cost and needs approximately the same labour input as the 'traditional' scove clamp. The technology certainly appears to be benefiting low-income producers; and
- consumers benefit from an increased supply of bricks.

The major problem with evaluating the appropriateness of the clamp lies in the lack of quantifiable information. Questions that cannot yet be answered include:

- exactly how much energy does the coal-fired clamp save compared with the scove clamp?
- how many trees are preserved as a result of switching to the coal-fired clamp?
- by how much are small-scale producers better off?

Training was carried out in several different ways. The number of producers who have adopted and who are proficient in the technology indicates that this strategy has been at least partially successful. Unfortunately, there is no way of comparing what would have happened if the project had adopted a different approach to training, such as running regular courses. The quality of the training is difficult to assess. Brickmakers trained by ITDG have, however, been used successfully to train others. This indicates that knowledge has been transferred and that it is being built upon. Follow-up and extension work were not undertaken in a methodical way. Although many visits were made, the results were not always recorded. It was not possible to make the desired number of follow-up visits or to visit other areas of the country due to the shortage of staff. Comprehensive documentation of extension work suffered for similar reasons.

Disseminating information had perhaps two major elements: dissemination to brickmakers within Zimbabwe, and dissemination of the project experience internationally. In Zimbabwe information flowed into and out of the project in 'random' ways:

- information sheets were supplied to enquirers;
- project staff spent many hours talking to interested parties;
- brickmakers told other brickmakers;
- ITDG presented its work at conferences and exhibitions; and
- institutional partners spread the word through their own extension and training work.

So, many methods of dissemination have been used in response to demand. Once again, though, there is no way of comparing what would have been the result of a planned and proactive programme. The fact is that, by adopting a mainly responsive approach, the project generated more than enough work for the staff and resources available.

In terms of disseminating the work of the project internationally, the project did not fully exploit its potential. The experience of the project, both in terms of developing an energy-efficient technology and transferring that technology, should have generated more articles and papers. Over the years, though, a number of attempts to communicate more widely were made. There have been some articles (Mutsambiwa, 1993; Tawodzera, 1994) and the project produced a video of its work. Nevertheless, dissemination could almost certainly have been better.

Monitoring and evaluation of the project could also have been improved upon. It could have more usefully served the needs of brickmakers, ITDG (both in Zimbabwe and internationally), policy and decision makers, and the international NGO community.

Marketing the bricks. The project staff has not needed to be involved with this too much, since neither the product nor the market has changed fundamentally. Where brickmakers

> **Box 6.2** Steve Murambidzi's brickworks: a two-way collaboration
>
> As a qualified computer technician and director of his own small software retailing company, Steve Murambidzi might be judged an unlikely brickmaker.
>
> 'Like everyone these days,' he says, 'I have to do what I can to make ends meet.'
>
> Working a very long day and dividing his time between his interests, Steve lives in Epworth, a settlement on the outskirts of Harare that started life as a squatter camp. His brickworks is also located in Epworth. At peak times Steve employs up to 30 people making bricks.
>
> 'Since we started working with ITDG,' he says, 'it's changed from a hand-to-mouth venture to a full-time, small-scale brick-moulding project. ITDG encouraged us to have foresight and to grow. We've improved product quality and increased sales. We've advertised our bricks in the newspaper – and even on the internet!'
>
> With the project's assistance, Steve has learned more about bookkeeping, business management and keeping records. He has drafted a business plan complete with a cash-flow forecast for expanding his operations and investing in brick-moulding machines. Steve sells his bricks to the local community and all over Harare. Demand still far outstrips supply: he sells everything he can make and his order book is full for months in advance.
>
> Steve's alliance with the project has benefited both sides: he has played an important role in research and development. The environmental impact of brickmaking has been investigated at his site, and this has resulted in recommendations for further research into alternative fuels from waste materials, coppicing trees and recycling. The energy use of a number of clamp designs using various fuels has also been monitored and recorded. The coal-fired clamp offered substantial energy and cost savings over wood or hybrid clamps. The substitution of boiler waste for coal further reduces costs. These days, however, Steve has to use what he can get.
>
> 'The fuel we use is determined by cost and availability. Sometimes coal is too expensive to buy enough for firing an entire clamp and we can't get boiler waste. Then we use the hybrid wood and coal clamp that we developed – even though wood is getting very scarce. But we'd rather use all coal or boiler waste because the quality of bricks is much better.'

have improved the quality of their product – upgrading from farm to common bricks – they appear to have had no problem in addressing a new market and obtaining a higher price. Similarly, there has been little need for consumer education: customers generally know the quality of bricks they require. Where no construction or material standards apply, consumers will buy the cheapest available bricks; in urban areas, however, people are prepared to buy bricks that are acceptable in that context.

From the outset, the project sought to have a good working relationship with the brickmaking community – the beneficiaries of the project – and in the main this goal was achieved. Brickmakers trust the technical advice of ITDG and project staff are constantly learning from brickmakers about socio-economic conditions and the marketplace in which they operate. In addition, many enquiries and requests for assistance continue to be received. ITDG make an organized attempt to record the number of enquiries. This is potentially a good indicator of the project's on-going impact.

Conclusions – but not the end of the story

Overall the experience of ITDG Zimbabwe with small-scale brickmakers has been positive. The coal-fired clamp technology is appropriate, effective and widely adopted; project staff members have a good relationship with the beneficiaries and a good understanding of the socio-economic and market forces that exist.

One reason for the success of coal-fired clamp technology is that it does not require great changes in brickmaking techniques, product marketing or construction practice. Using the coal-fired clamp simply enables brickmakers to save fuel and hence production costs.

The process of technology transfer has been a little chaotic. Nevertheless, there are positive lessons that have been learned for the project in Latin America. Flexibility within a planned strategy would appear to have been a key element of success. On the other hand, a shortage of staff and resources has been a severe constraint.

Perhaps the area where feedback is least positive is in the rigorous documentation and thence dissemination of information. This is particularly apparent with respect to information on energy usage, fuel consumption and burning efficiency.

7
THE START OF THE PROJECT IN PERU – GATHERING DATA ON ENERGY EFFICIENCY

TAKING INTO ACCOUNT the experience in Zimbabwe, the project in Peru had at the outset to find ways to answer four seemingly simple, related questions:

- How efficient are the traditional, wood-fired, Scotch kilns used in Peru and the methods of firing employed in Ecuador?
- How efficient are the layered, coal-fired clamp and other technologies promoted by the project?
- What is the improvement in efficiency and cost saving brickmakers can expect if they adopt these technologies?
- How might the technologies be adapted to suit local needs and conditions better?

This chapter describes the design principles of technology transfer that became known in the project as the 'Ten Commandments', and that were applied in the project in Peru and Ecuador. It then goes on to describe a method for measuring the specific firing energy use of a kiln, measured in joules per kilogram of fired brick, and how this figure must be qualified by temperature measurements indicating how well fired the bricks have been. These measures can then be used to answer the first three of the above questions. The fourth question, describing how brickmakers in Peru and Ecuador have attempted to modify introduced and local firing technologies, is answered in the next chapter.

Designing the technology transfer project in Peru

The project staff in Peru were aware of the theories, opinions and case studies on technology transfer. They carefully considered the issue of appropriate technology in the context of Latin American brickmakers and the coal-fired clamp. They wished to adopt a participatory approach and the elements of PTD. Combining these considerations with the earlier experience of working with brickmakers in Zimbabwe, project staff evolved their own trial technology transfer methodology. None of the writers and commentators on the subject had outlined a step-by-step methodology, and even the case studies available were mainly retrospective examinations of how a technology transfer process had fared rather than plans for implementing such a transfer. In general, the NGO sector did not appear to be documenting its experience very well. Perhaps that was because there was not much rigorous experience to record.

It has to be remembered that NGOs spend most of their time dealing with the poorest producers who have no other path to effect technical change: there are no equipment sales staff or commercial technology consultants queuing up to offer them advice. Industrial-scale enterprises in the developing world have evolved to the stage where they are able to articulate their need for new technology. They are *ready* to develop technologically and capable of seeking out the technology that best suits their needs. Project staff in Peru bore in mind that this might not be the case with small-scale brickmakers – or at least that the process was bound to be very different. Small-scale brickmakers often express enthusiasm for mechanization without necessarily thinking through the implications this would have for accessing capital, labour requirements, developing new markets and so on. Part of the participatory process might involve project staff assisting brickmakers to consider such ramifications and to perhaps better define their technological needs.

There was, however, marked agreement on one point in all the theories of technology transfer and PTD, whether the experience concerned the commercial or the non-profit sector: research and development *located in the developing country* was seen as a critical factor for success. In addition, there was general agreement on the need for beneficiary involvement, training, information dissemination, marketing and consumer education.

Research and development must take place locally. The project staff therefore decided to carry out research and development of energy-efficient brick-firing technologies in Peru and later in Ecuador. They would use and develop local knowledge and skills, and the technologies examined would be tested for their appropriateness in all aspects. This testing would be carried out with the involvement of local organizations and using their facilities, and it would therefore fulfil another important objective: to build the capacity of local institutions.

Building up local training capacities. Training was obviously a key element to which the project would pay particular attention. The project staff began to ask themselves important questions: Could they design the training element of the project so that it could eventually run independently of the project? For such training to continue to be available after external agencies like ITDG had withdrawn, it would probably be necessary for a training organization to charge fees to the brickmakers to cover their costs. Which organizations would be appropriate allies? Did small-scale brickmakers have the resources, capital and, perhaps more importantly, the expertise necessary to choose the techniques that would suit them? Were they sufficiently educated to take full advantage of training? What form should training take? Could brickmakers afford to pay training fees? Could they even afford to dedicate the time? It turned out that although the project in Peru and Ecuador was to make a number of institutional allies, none of these was able to provide appropriate training that the brickmakers would have been willing to pay for (see Chapter 9).

How to disseminate information on successful technologies? Access to information that would enable a brickmaker to choose from among a number of techniques one that would suit his or her needs was a crucial factor. This is particularly difficult where producers live in scattered rural, peri-urban and urban-fringe areas, and are never normally likely to gather together to view demonstrations of different brickmaking techniques. Given ITDG's very limited resources, widespread dissemination would have to be via assimilation by individual brickmakers and imitation of such local brickmakers. Information had to be pitched at a level appropriate to the recipients' level of education. The information had to be accessible and comprehensible. Exchange visits by brickmakers from different areas proved to be a valuable method of disseminating training on new technologies, such as the oil-fired burner (see

> **Box 7.1 Even the best make mistakes**
>
> Since independence from Great Britain in 1966, Botswana, which began life as a poor nation, has grown into an African success story. It is one of the least corrupt countries in sub-Saharan Africa, boasts a stable democracy based on the traditional *kgotla* system, has little or no debt and, in 1999, had one of the fastest growing economies in the world, with a real growth rate of 9 per cent.
>
> At a public meeting in Serowe, Botswana, shortly after Quett Masire became president, the venerable politician said in his speech that he would take complaints about lack of consultation in policy making very seriously. He then went on to announce from a podium, before this audience of three or four hundred, that that was exactly what he was doing at that moment – consulting the people!

Chapter 8). In the longer term, ITDG staff agreed, education would be needed to back up training and information dissemination.

Understanding the market, and the producers. ITDG grouped the elements of marketing and consumer education together. For low-income consumers, price was the overriding consideration when making a purchasing choice. Other consumers might be more influenced by quality, availability, speed of delivery or other factors. For example, in urban areas where building standards existed or where multi-storey buildings were to be constructed, the customers would demand a certain strength and appearance from the bricks they purchased. They would also require a guarantee of timely delivery. Once again, ITDG staff began to phrase the questions they would have to address. They would need to carry out a market study to assess both the existing situation and the changes that project implementation might imply. Was there genuine scope for improving the lot of small-scale brickmakers? Who were their customers? And who were their competitors?

The market study revealed some of the differences between Zimbabwe and Peru. The market for small-scale brickmakers in Zimbabwe is generally for farm bricks; but brickmakers were making efforts to upgrade their product towards the standard required for common bricks. In the vast majority of cases, production is not diversified in Zimbabwe: small-scale, informal-sector brickmakers do not make roof or floor tiles. In Peru, and to a lesser extent Ecuador, however, very small-scale, low-tech brickmaking enterprises commonly engage in making roof tiles and roof bricks also, a product for which there is a ready market.

There were also a number of subtle organizational differences between brickmaking enterprises in Zimbabwe and Peru and Ecuador, which would have an effect on the project. In Zimbabwe the model for peri-urban brickmakers, engaging in commercial scale production to supply the town market, is most likely to be an entrepreneur employing local labour on an informal basis. The entrepreneur is most likely not to be a brickmaker and will not often be present on the site. He – most are men – will probably have other irons in the fire (as in the case of Steve Murambidzi, Box 6.2). This absentee entrepreneur model has mainly replaced the, largely discredited, quasi co-operative structures that burgeoned briefly in the years immediately following independence in 1980, when Zimbabwe briefly fostered socialist aspirations.

Photo 7.1 Mixing soils prior to moulding the bricks involves hard physical labour on the part of women, men and cattle

ESPOCH, Ecuador

In Peru and Ecuador, more often than not the head of a brickmaking enterprise is a brickmaker, a man or woman who is present and involved in production. These brickmakers appear more readily disposed to the formation of associations through which they can pursue collective interests, though they invariably maintain control of their own enterprises. They might, therefore, be more likely to market collectively or seek government or NGO assistance together.

The involvement of the recipients of technology is vital. ITDG in Peru had to design their intervention to address the specific problems of the brickmakers. They had to consult with brickmakers to determine the most appropriate form of assistance. The project had to be designed to allow for clear communication and provide accessible channels for feedback. Project goals and constraints had to be clear to the brickmakers from the outset. Brickmakers had to have an intellectual and emotional stake in the project. Their commitment to it was essential and they had to be motivated to succeed. It was deemed vital that brickmakers felt like allies rather than pawns. Brickmakers had to be encouraged to articulate their problems, aspirations and priorities. And ITDG had to listen. Eventually, brickmakers might need an appropriate forum for expressing their common concerns to a wider, influential audience.

Impact of the intervention. ITDG had to ask questions beyond those that would interest the commercial sector: they had to assess how any benefits from the project were likely to be distributed. Would the boss of a brickyard get rich while nothing improved for the workforce or their dependents? Could ITDG do anything about this? Was it beyond the scope of the intervention? How could the project ensure that employees were not exploited; that they had satisfactory wages and working conditions? Apart from the brickmakers, ITDG had to ask

> **Box 7.2 Cold comfort – an unexpected project impact**
>
> Things are often not as straightforward as they appear. An ITDG project in Sri Lanka looked at the production and marketing of a variety of fruits and vegetables. Obviously, the crops were harvested at specific times in the season and at these times there was a glut: the price fell, fruit and vegetables rotted, and producers lost out. It seemed clear that they needed help. The project assisted a grower to install a cold store, which allowed him to spread selling his produce over an extended period. It worked. For that grower the price fluctuations were ironed out and he could supply certain sorts of produce when no one else was able to. He was even able to raise his prices. Wonderful news; a great project, surely?
>
> Maybe not.
>
> Before the experience was replicated on a wider scale, project staff carried out a survey of the customers who normally bought the produce. They found that a significant percentage were very poor people who could only afford to buy when prices were reduced by a glut on the market. The poorest even had to wait until the vegetables were starting to rot and they could obtain them free. For the majority, the introduction of cold stores on a wide scale would mean they would be able to eat hardly any fruit and vegetables. Ultimately, it would mean that the producers would not have enough customers for their stored vegetables at higher prices. They would probably have to sell them off cheaply. Given the market and the socio-economic conditions, implementing the cold storage project would have been a wasted investment.
>
> Back to the drawing board.

whether the project would have an adverse affect on any individual or group. Would it lead to practices that were less environmentally damaging?

There were also specific impact questions relating to the transfer of energy-efficient techniques and practices. At first sight the premise seemed simple: saving energy used in brick production would save the producers money and, in global terms, conserve scarce energy resources. But what would be the 'knock-on' effects of such a seemingly straightforward intervention? If producers saved money, would they pass on the benefits to their customers? Would they sell bricks more cheaply, thus benefiting low-income consumers? Or would they hold their prices and thus increase their profit margin, thereby benefiting themselves – the low-income producers? What if, as a marketing strategy, they actually increased the price of their bricks, claiming a better-quality product?

A change of fuel – for example from wood to coal – might also have unforeseen implications. Who supplied the wood at present? Would these suppliers – and their families – suffer as a result of the switch to coal? From the brickmakers' point of view, could they afford, individually or collectively, to make the investment to stockpile sufficient reserves of coal?

Any technology change is likely to have negative impacts as well as positive ones, and often the latter outweigh the former and the project goes ahead. This is not always the case, however, as seen in the cold store project described in Box 7.2. Here, not only was the negative impact experienced by the poorest customers, but since it was difficult to reach markets other than these poor customers, even the hoped-for boost in income for the producer was unlikely to be realized. Where negative impacts are experienced by the disadvantaged – by poorer sections of the community, by women, or by labourers – this may run up against the mission statement of non-profit organizations dedicated to helping the poor.

While considering all these initial questions, the project staff was well aware that other, even more complex, considerations would undoubtedly arise as the project proceeded.

The draft technology transfer plan

It was important that the project was not over-ambitious. Technological change and the development of skills took time, and there would probably be an innate conservatism and resistance to change in both producer and consumer attitudes. In the initial stages, therefore, the project would work with a few selected brickmakers, in the hope that if a small number of brickmakers were seen to be adopting energy-efficient practices and perhaps alternative fuels, and becoming better off, then this would have a knock-on effect. Others would imitate their methods to emulate their success. Working with a select group would, in the long term, be more effective than spreading the project's limited resources too thinly.

A flexibility of approach and attitude on the part of project staff was essential. Although the available technologies for brickmaking were unlikely to change radically, prevailing socio-economic and politico-cultural conditions *were* subject to fluctuation. Additionally, staff had to bear in mind that, ultimately, the coal-fired clamp and other known practices might not be the right technologies in this context; other innovations might be more valid. In some cases no change at all to current practice might be justified. The indigenously evolved technology might already be the best choice. This could apply particularly to very small-scale, part-time brickmakers situated in remote areas with a very limited market and no access to credit. At all times, staff remained aware that technology was a package.

The brickmakers in La Huaca, Peru, for example, were already experimenting with incorporating rice husk into their kilns to reduce fuel costs before the project intervened (see Chapter 8). These were people who were able to innovate already: what they needed was for the range of their technological options to be widened, and to be assisted in monitoring the savings of different technologies.

The project team elected to adopt the following principles and practices of technology transfer and participation, which were wryly dubbed the 'Ten Commandments'. At worst these principles would serve to guide the project in its initial stages; at best they would be the firm basis for a step-by-step technology transfer methodology. Though the commandments refer to brickmaking in Latin America, they may well be adaptable for technological interventions in other sectors in other regions.

The Ten Commandments of technology transfer and participation

1 Staff should have a thorough understanding of the brick industry, the market and the socio-economic situation of brickmakers.
2 Staff should be competent and confident in the technology being promoted. They should know and be able to explain the potential benefits of adopting such technology.
3 Brickmakers should be involved in the planning and implementation of the project from the outset. They should be encouraged to be active allies with a voice in decision making and project development.

4   Research and development should be undertaken in Peru and Ecuador. This R&D might involve co-operation with local universities, colleges and appropriate institutions. If on-site research involved any financial risk for brickmakers, the project should be prepared to cover the cost.
5   A short-term project methodology should be evolved and tested. This should involve a trial period of working intensively with a small number of brickmakers to establish the technology as viable.
6   Training is the key to successful adoption of energy-efficient brickmaking practices. Appropriate training methodologies should be selected to meet brickmakers' requirements.
7   Information on energy-efficient practices should be made available to brickmakers and their allies. This information should be in an accessible form, accurate and objective.
8   The project should be faithfully documented, including failures as well as successes, and the reasons for them, with a view to learning lessons and ultimately disseminating the methodology.
9   Beneficiaries and any individual or group potentially adversely affected as a result of the project should be closely monitored.
10 The likely effect on the environment should be monitored.

The importance of measuring energy efficiency

The first big step for the ITDG project in Peru was a technical one. The crux of the intervention was technologies that saved energy and fuel costs, thereby benefiting the environment. Therefore, these technologies needed to be *proved* more energy efficient and financially viable. The non-technical reader whose interest is more in participatory methodologies than brickmaking will, hopefully, find the following information of interest. It illustrates clearly how dealing with appropriate technology and the small-scale production sector does not excuse professionals from scientific rigour. Indeed, the rigorous application of science and an imaginative and innovative approach is often more critical (see Box 7.3).

> **Box 7.3   Some great expectations**
>
> If (the engineer) has chosen to work with small-scale producers and appropriate technologies, then he or she has probably chosen the most difficult field of engineering. This engineer cannot throw armfuls of money at a technical problem. He or she does not have the backing of a huge corporation. In addition, he or she cannot consider solutions that require substantial investment on the part of small-scale producers. No matter how short the pay-back time, they generally do not have the spare capital to invest. Most often they cannot borrow money either, being excluded from the formal process of bank loans. Furthermore, our appropriate technology engineer has to pay closer attention to the social and environmental impact of an intervention than does any other engineer. 'The environmental crisis the world faces today is not due to one massive technical mistake but to millions of seemingly trivial miscalculations' (Floorman, 1994).

The project aimed to introduce technologies that would increase the efficiency of firing bricks. Ultimately these technologies would either save brickmakers money or at least offer them a viable fuel alternative to increasingly scarce firewood. The problem the project team faced was how to compare the efficiency of brickmaking processes so that the technologies introduced could be evaluated. The project had to build on and develop the experience from Zimbabwe. Early attempts to compare efficiency immediately revealed a number of problems, however. What at first seemed a relatively simple matter was in fact fraught with difficulties. A number of factors affect the energy used in brickmaking, and in order to compare the efficiency of different processes they all have to be taken into account.

Factors that affect energy use in brick firing

- The nature of the clay: its refractoriness or resistance to change by heat; its intrinsic calorific value – the clay itself may contain organic matter that acts as a fuel; the presence of fluxes – substances that aid vitrification. In England, for example, yellow 'Fletton' bricks require as little as one-third of the energy to fire them as heavy clay bricks.
- The moisture content of 'green' bricks going into the kiln. Green bricks are nominally dry but unfired bricks. But in many cases they are not completely dry and energy is used to evaporate the water they still contain.
- The type of fuel used, its calorific value, moisture content and distribution in the clamp or kiln.
- The design of the clamp or kiln.
- The skill of the brickmakers controlling the burning process.
- The climatic conditions: ambient temperature, wind direction and strength, and so on.

What casual data recording does not include

Information on the energy required for brickmaking had largely been collected and presented in a random fashion. Often the data were useless in scientific terms because they had not taken into account the factors listed above. Indeed, much of the available data ignored even more fundamental considerations.

For example, a report might state that a certain brickmaker used 12 cords of wood to fire 30 000 bricks. This information was probably very useful to that particular brickmaker. But what does it tell a wider audience about the energy used? We do not know the type or weight of the wood. A cord is a volumetric measure that *can* indicate a stack measuring 2 m × 1 m × 1 m. But without knowing the type and average diameter of the wood, it is impossible to make anything other than a guess at its weight. Even then we have no indication of its moisture content or calorific value. Textbook values for the energy of fuels, especially wood, can only be a rough guide.

And what about the bricks? We do not know the soil type – nothing about refractoriness or the presence of fluxes, nothing about the energy required for vitrification. We do not know the size of the bricks. In some regions of Peru they make giant bricks, which are twice the mass of the 'common' bricks produced in Zimbabwe: these bigger, heavier bricks obviously take more energy to fire. We do not know the moisture content of the bricks when firing started, and hence how much energy was used simply to finish drying them. We do not know the colour, mass, density, porosity, hardness or compressive strength of the fired brick. That is, we have no indication of how well-fired they are. The bricks could be anything from 'smoked' (not vitrified at all) to 'hard-burned'. We have no idea of the firing temperature reached nor for how long it was maintained. Neither do we know how much energy the bricks 'absorbed', that is, how much heat was used effectively and how much was wasted to the atmosphere.

In short, we know next to nothing, and certainly nothing useful to aid comparison with other firings.

Methodology for measuring the energy used to fire clay bricks

The first step for the project in Peru was therefore to develop an international standard for the collection, analysis and presentation of data on brick firing. One difficulty was instantly apparent: facilities and services available around the world would differ. Therefore, the standard would have to be 'an operable field standard', minimizing specialist services and costs. In order that the standard could be widely adopted, the collection, analysis and presentation of data should involve as little extra work as possible for brickmakers. After considerable effort and consultation, the project evolved the following methodology.

Gathering data

The first points that researchers should note are obvious: the name of the brickmaker, the location and/or address of the brickworks, the date and duration of firing. This may be obvious, but it is all too often forgotten. Then there are six essential pieces of information to record.

1 *The number of green bricks going into the clamp or kiln.*
2 *The average 'wet mass' of a green brick.* Prior to kiln construction, a number of green bricks, representative of the typical brick, should be selected and weighed. The number in the sample should be one brick per thousand bricks fired in the kiln. For smaller kilns, the minimum number of bricks should be 24. Once bricks have been weighed, the average 'wet mass' can be calculated.
3 *The average 'dry mass' of a brick.* It is necessary to ascertain the moisture content of the green bricks and, ultimately, the energy needed just to dry them. Hence, the next step is to dry the selected sample bricks and obtain a figure for the average 'dry mass'. Bricks can be dried in a conventional oven or a simple field oven can be made. The important thing is to maintain the bricks at about 80°C and to weigh them periodically. The average dry mass can be calculated when no further weight loss is noted. Some experts believe it is preferable to dry bricks at 105°C.
4 *The total mass of fuel(s) used in the kiln.* Fuels like coal and boiler waste are easier to weigh than wood. If a fuel such as sawdust or pulverized fly ash is mixed into the body of bricks, then its mass must also be determined.
5 *The calorific value(s) of the fuel(s).* A small sample of *each* fuel used should be taken promptly to test for calorific value and, ideally, moisture content. If facilities to test calorific value are unavailable or unaffordable, the best that can be done is to approach a university or other research establishment for the best local data. If a value is found in this way, it should *not* be recorded as the specific calorific value of the fuel, and a note should be made of the information source. (Life and science are seldom easy. In one rural clamp monitored by the project team in Zimbabwe, the brickmakers used 14 different types of indigenous hardwood in different quantities.)
6 *The average mass of a fired brick.* Once the kiln is fired and cooled, a representative sample of bricks – the same number as before – should be weighed and the average 'fired mass'

calculated. The bricks do not have to be the same ones that were weighed when green or dry.

A larger-than-expected difference between the dry and fired masses of bricks can result from the fact that the clay contains naturally occurring organic matter, which burns off when bricks are fired. If this is the case, the soil should be tested to ascertain its calorific value as this will obviously make a difference to the energy calculations. Alternatively, a large difference in weights may indicate an error in weighing bricks and this should be the first thing to be checked. (In the case of all the completed monitoring forms presented in this book, it is suspected that the brickmaking soil contained a percentage of organic matter.)

There is additional qualifying information that can increase understanding and facilitate better comparison with other processes. It may be worth noting the kiln dimensions and calculating its volume; then the total volume of green bricks can be obtained by measuring the dimensions of the sample bricks, calculating an average volume and multiplying by the number of bricks in the kiln. The 'void volume' – which relates to airflow through the kiln – can then be calculated by subtracting the volume of bricks from that of the kiln. The number and orientation of the bricks in each layer could be noted and the 'per layer' void spacing calculated accordingly. This information will allow brickmakers to replicate a successful kiln better or to modify a less successful one. Other points that it could be useful to note include:

- The kiln firing time – how long the kiln burns after external ignition.
- The 'active' firing time – how long the kiln is actually being 'fed' fuel.
- The percentage of fuel used 'externally' – in firing tunnels or grates.
- The 'soak' time – how long it takes the kiln to cool after the fuel is burned.
- The climatic conditions – wind direction, its nature and strength, the orientation of the kiln, rainfall and ambient temperature.
- The altitude – at high altitude fuel may burn at a lower temperature and hence bricks either take longer to fire or do not vitrify as well; the practical effect might be that marginally more fuel is required.

Analysing the data

The qualifying data mentioned, along with any other relevant observations, can be useful. However, specific firing energy, or the energy used per kilogram of brick produced, is calculated using only the six essential points listed. This method is best demonstrated using an example (see Box 7.4) based on the actual data obtained from a case study conducted by the project in January 1998.

Qualifying the results

From experience, the project researchers realized when they analysed the data from the case study that the specific firing energy of this clamp was comparatively low. With the technology used, and in this specific situation, they would have expected a typical specific firing energy of 2000 kJ/kg or more. So what could they derive from this result? They could speculate that the burning process was either very efficient or that the bricks were under-fired, but they could not draw any firm conclusions. They needed to know more. And they knew where to look. They had not taken into account those factors which were earlier listed as vital in any analysis: the soil type and what temperature it vitrified at; and how well the bricks were fired – how

Box 7.4 Calculating the specific firing energy for a brickmaking method

The first step is to calculate the moisture content of the bricks. The average wet mass of a green brick was measured as 3.65 kg, and the dry mass was 3.25 kg, so the moisture content is 0.40 kg per brick. Next, the energy needed to evaporate this moisture is calculated. The kiln held 14 817 green bricks, so

$$\text{Total mass of water to be evaporated} = 14\,817 \times 0.40 = 5927 \text{ kg}$$

The heat of vaporization of water is taken as 2255 kJ/kg and the energy required to raise the water to vaporization temperature is estimated at 336 kJ/kg. Adding these figures

$$\text{The specific drying energy} = 2591 \text{ kJ/kg}$$

The total energy required to dry the bricks can now be calculated:

$$\text{Drying energy} = \text{specific drying energy} \times \text{total moisture content}$$
$$= 2591 \times 5927$$
$$= 15\,356\,857 \text{ kJ}$$

The fuel used was 4680 kg of boiler waste – partly burned coal from the local power station. The boiler waste had a net calorific value of 19 180 kJ/kg.

$$\text{Total energy} = \text{mass of fuel} \times \text{net calorific value}$$
$$= 4680 \times 19\,180$$
$$= 89\,762\,400 \text{ kJ}$$

In this example, drying energy accounts for about 17 per cent of the total energy used. In many cases it could be more.

$$\text{Firing energy} = \text{total energy} - \text{drying energy}$$
$$= 89\,762\,400 - 15\,356\,857$$
$$= 74\,405\,543 \text{ kJ}$$

The average fired mass of a brick was measured and found to be 3.10 kg.

$$\text{Total mass of fired bricks} = \text{average mass of a fired brick} \times \text{number of fired bricks}$$
$$= 3.10 \times 14\,817$$
$$= 45\,933 \text{ kg}$$

It is now possible to calculate the specific firing energy:

$$\text{Specific firing energy} = \frac{\text{Firing energy}}{\text{Total mass of fired bricks}}$$
$$= \frac{74\,405\,543}{45\,933}$$
$$= 1620 \text{ kJ/kg}$$

much heat was used effectively, and how much was lost. So, to make sense of results, the standard must record information that *qualifies* the result obtained for the specific energy used.

First, it is necessary to know the forming process used, as this will make a difference to the energy required to burn the bricks. For example, bricks made using a semi-dry pressing process typically need a higher temperature to vitrify than hand-moulded bricks.

Then, the project team decided to categorize soil types as 'high temperature' (vitrifying above 1000°C), 'medium temperature' (vitrifying between 950 and 1000°C) or 'low temperature' (vitrifying below 950°C). Unfortunately, being able to do this requires access to a laboratory or potter's kiln that will achieve such temperatures accurately. The procedure is to mould cones of soil, dry them, fire them and note the temperature at which they begin to bend or slump.

At the other end of the process, some indication was required of how well the bricks were fired, or how much useful work the firing process had done. Measures that the project team considered in developing a standard included the following properties of fired bricks: density, porosity, hardness, compressive strength and so on. It was decided, however, that these brick properties would also be determined by factors other than firing, such as the type of clay used (see Chapter 2 for a list of other factors). Clearly, the final quality of the brick is of paramount interest to the brickmaker, but at this stage the main concern of the project team was the minimum energy required to produce an adequately fired, and not under-fired, brick.

The effectiveness of a brick-firing process is a function of time and temperature. The relationship is not linear. As an extreme example, bricks fired at 1000°C for 10 hours would not be the same as bricks fired at 100°C for 100 hours! Temperature could be measured against time at various places in the kiln. This information could then be presented graphically to qualify the specific energy result in each case. The drawback with this is that the thermometry required to measure up to 1200°C at a number of locations is expensive.

A more affordable and useful alternative is 'Bullers bars', which are commonly used in the ceramics industry. Bullers bars melt or sag when they have reached a certain temperature and that temperature has been maintained for a sustained period. They can, therefore, be used to indicate a particular condition relating to time and temperature. Bullers bars come in a range, deforming in conditions nominally related to temperatures from 590 to 1525°C ('nominally', because if you exposed a certain bar to its designated temperature for only one second, it would not sag: it takes longer than an instantaneous exposure). The bars are 57 mm long, and a set of four, which deform at different nominal temperatures, is usually placed in a refractory stand to form a 'thermoscope'. These are then placed strategically in the clamp or kiln. After the kiln has cooled down, the thermoscopes removed from different parts of the kiln will reveal which number of bar sagged in which part of the kiln, and therefore what was the maximum temperature reached in different parts of the kiln. The number can then be compared directly to the index for any other kiln. This rules out other factors that might confuse comparison.

Returning to the example in Box 7.4, the bricks were formed by semi-dry pressing, though in this instance the clay was noticeably too wet to be ideal for that process. Cones of clay were fired in a potter's kiln and did not bend or slump at 1000°C. Thus, the soil was categorized as 'high temperature'. Eighteen thermoscopes were positioned and recovered. They contained bars 11, 13, 15 and 17, corresponding to nominal temperatures of 845, 890, 940 and 990°C, respectively. Nine of the thermoscopes, recovered from the upper portion of the kiln, were unaffected – none of the bars had sagged. Of the other nine, recovered from the lower portion of the kiln, four showed bar 13 just sagging while bar 11 had virtually melted. The other five thermoscopes showed bar 11 sagging, but bar 13 unaffected.

It would be possible to say the kiln in its entirety had an average heat–work index corresponding to a Bullers bar number of 12. That is evidently not the full story, however. What the result shows most clearly is that (a) there was uneven heat distribution through the

Photo 7.2 A set of Bullers bars in their stand is placed in the kiln as it is being loaded

Kelvin Mason/ITDG

kiln and (b) that the vitrification temperature needed for this particular soil was never achieved. The specific energy used in this kiln can be concluded to be low because the bricks were under-fired. Compressive strength and water absorption tests on the bricks confirmed this conclusion.

On the other hand, the observation recorded at the time of firing stated that 95 per cent of the bricks were well-fired: a percentage that is so high that it is often taken to indicate that the batch was over-fired in terms of the balance between minimizing energy wastage and minimizing brick wastage. In many of the energy-monitoring forms derived from the project in Peru and Ecuador, the Bullers bars indicated that vitrification temperature had not been reached, but that the level of under-fired bricks was very low (see Appendix 2 for further examples). The reason for this contradiction between the casual observation and the temperature measurement is that 'well-fired' to the casual observer probably means '95 per cent look as good as usual' – and the 'usual' artisanal brick does not reach its vitrification temperature all over the world (see Chapter 2). Clearly, there is a margin below the vitrification temperature for a particular soil that if the clamp reaches this lower temperature most of the bricks will be strong enough for the purposes to which they are going to be put.

Presenting the findings

Ultimately then, the results of a brick-firing process should include:

- the specific energy used;
- the moulding process;
- the soil temperature category;
- the average Bullers bar number;

- a statement of qualifying information, including a description of heat distribution in the kiln; and
- a record of the number of unusable or unsuitable bricks: those that are under-fired, weak and/or broken; and those that are over-fired, discoloured and/or distorted.

Feedback

The project team believes that this standard methodology represents a significant advance in comparing energy efficiency in brickmaking. It should prove extremely useful to brickmakers and those working with them. It is one of the most significant technical outputs of the project, and has been adopted by organizations such as TERI (Tata Energy Research Institute), New Delhi. The standard methodology is published as a practical technical brief (Mason, 2000) for fieldworkers, complete with a photocopiable blank monitoring form. A field worker from TERI wrote that 'In the absence of a standard protocol for collecting information on energy use, a lot of confusion is caused when we try to compare information collected by others with our information on specific energy consumption .... The technical brief is useful because it provides a methodology that can be used in the field in developing countries' (S. Maithel, personal communication).

On a cautionary note, although this method of measuring kiln performance has been designed to be as simple as possible, it does take a number of people to monitor the inputs to a kiln effectively. It has been found that at least four competent staff members are needed. Typically, one would record the number, orientation and placing of bricks. Another would weigh and note the fuel used. A third would carry out brick drying and weighing tests. And the last would co-ordinate the collection of data, note qualifying and additional information, and liaise with brickmakers.

In theory, small-scale brickmakers who wished to monitor their kilns could make the required measurements – they did in fact carry out most of that work – and they could learn to make the calculations to determine specific firing energy. It is unlikely that they would want to do this for every firing, however: once they had discovered a suitable process, they would only want to monitor kiln performance if they intended to make changes or if the process was seen – in declining brick quality – to be going wrong. Keeping a basic record of how much fuel was used, and any changes in practice, was something that was advocated for routine monitoring, however, and this was noted in the 'Ten Rules' for cost-effective brickmaking evolved by the brickmakers in La Huaca (see Chapter 8). The full energy-monitoring methodology is more likely to be employed by engineers working with brickmakers or researchers. Of course, a technically educated brickmaker would be able to make good use of the methodology in production and quality control; but most brickmakers, even sophisticated high-tech operations in 'the West', use experience as their principal technical guideline, commenting that 'the process has worked all right for years'.

It is also important to realize that there is a marked difference between carrying out scientific experiments in a sheltered academic environment and doing research in the field. Box 7.5 describes the author's attempts to measure the specific energy use of a kiln in which coal is being substituted for fuelwood, and the difficulties that arise and almost wreck the exercise. No matter how well prepared one is, something – probably several things – will go wrong.

Energy consumption of brick firing processes: Form 07		
NAME OF PRODUCER Víctor Carmen	LOCATION /ADDRESS La Huaca, Paita, Piura, Perú	DATES OF FIRING Start 20 June 1997 08:00 Finish 21 June 1997 07:00 wood
TYPE OF CLAMP/KILN 3 tunnel, Scotch kiln Size: 3.40 m × 3 m × 3 m	TYPE OF FUEL Algarrobo wood	MASS OF FUEL USED (kg) 5100
CALORIFIC VALUE (kJ/kg) (i) Algarrobo Gross = 17 555 kJ/kg Net = 16 310 kJ/kg Moisture cont. = 10%	NO. OF GREEN BRICKS 6350	MASS OF BRICKS wet = 4.20 kg dry = 4.13 kg fired = 3.80 kg
BRICK MOISTURE CONTENT 1.67 (%)	METHOD OF FORMING Slop moulding	WEATHER CONDITIONS Dry, hot; gusting light wind
CALCULATION OF KILN EFFICIENCY Mass of green brick = 26 670 kg Total moisture content = 445 kg Firing energy = 82 029 300 kJ Mass of fired brick = 24 130 kg Specific firing energy = 3.40 MJ/kg		QUALIFYING INFORMATION (i) Vitrification temp. = 1150°C Category of soil = high temp. (iii) Max. temp. = 970°C (iv) Firing time = 23 h
COMMENTS 95% well-fired, 5% under-fired and broken. Data reading taken on site by Mario Jara. Signature, date, organization and contact address: Emilio Mayorga/Teodoro Sánchez, 31 October 1997. ITDG-Perú, Lima.		

Figure 7.1   An example of a monitoring form for a traditional Peruvian Scotch kiln

Note: (1) The calculations in this form are based on the method of Box 7.4.

(2) This firing, among others, employed Bullers rings rather than Bullers bars. Bars were later found to be preferable.

Photo 7.3 A kiln is loaded with coal and bricks

Emilio Mayorga/ITDG

Box 7.5 The diary of a fieldworker

When the four of us arrived at La Huaca neither of our two main contacts among the brickmakers was present. Nevertheless, with some difficulty we located the kiln to which the coal had been delivered. The first snag: the coal was in dust not lumps – totally unsuitable for the type of firing we'd planned. I discarded all the work I had done over the past week. Did we know the calorific value of this coal dust or of the 'algorrobo' fuelwood that it was supposed to replace? Well, not exactly.

We proceeded to take measurements of the kiln and the bricks to estimate capacity. We questioned the brickmakers who were present about how many bricks it would contain and its fuelwood consumption. We prepared a plan to fire this kiln with a mixture of coal dust and wood. We made educated guesses about quantities and proportions. After a couple of hours we had all the information we could gather and a new plan.

On the way back to Piura, the town where we were lodging, we stopped off at the community centre in La Huaca to find that our two co-operators had been waiting for us there all day, with one of our colleagues, Miguel. Hadn't we made the arrangement clear? *Que sera*. After exchanging apologies, we briefly discussed what we'd ascertained in the afternoon. The brickmakers nodded agreement. That night, I stayed up late to work on the plan for building and firing the kiln the following day.

When we arrived in La Huaca in the morning we were shown the kiln that the brickmakers wanted to fire: a different kiln. It was much smaller than the one we'd documented, and the size of bricks they wished to burn was much bigger. I discarded the work of the previous day and began to make fresh calculations. We asked the brickmakers what was the capacity, to which they first responded eight, then seven, then seven and a half thousand bricks. And how much fuelwood did they normally use for this many bricks? After some debate, they decided it was around 600 poles. And the weight of a pole? 11 kg, more or less. We'd found general figures for the calorific value of algorrobo wood and coal dust and they would have to do. We made more calculations.

It transpired that the capacity of the kiln was only around 6000 bricks and the average weight of a pole is 8.5 kg. Later in the day, the brickmakers decided that actually they used more like 550 poles. I made a mental note never to trust hearsay – even when it comes straight from the horse's mouth. Our calculations would only have been thrown out by a factor of about 1.5!

We decided to replace half the normal mass of fuelwood used with its calorific equivalent in coal dust sprinkled between each layer of bricks. Though none of us had experience with this type of hybrid kiln, we had little choice. In the circumstances, it was not possible to use any of the proven technologies that we knew about. So we pressed on: weighed the coal into piles; counted the bricks going into the kiln; calculated the coal per layer; estimated the distribution; and weighed the fuelwood. We four and two assistants were fully employed trying frantically to keep up with the process.

Then, a brickmaker from Ayacucho, named Cesar, turned up. Cesar had experience of burning clamps with coal dust – 20 per cent substitution – and could pass on his experience to the brickmakers in La Huaca. We agreed with him that we'd carry on building the kiln with 50 per cent substitution of coal for wood, though Cesar was not very happy with this. All morning he argued that we were using too much coal. Then, as the kiln was almost complete, Cesar informed us that he did not think we were using enough coal!

When we'd finished building the kiln, at about three in the afternoon, we found we were pretty close to our target of 50 per cent energy replacement. I was in the process of explaining that, with the coal in layers, the firing of this kiln should be much slower and more constant than the normal bottom-fired kiln the brickmakers were used to, when Cesar and his friends stuffed the firing tunnels full of wood, poured on a liberal quantity of kerosene and started a blaze to rival the fires of hell.

Too tired to remonstrate, I built a small field oven to dry some green bricks and obtain the exact dry weight. Behind me, the kiln blazed furiously, igniting each layer of coal simultaneously, I imagined, and defeating the point of having it in layers. My smoky little five-brick kiln greatly amused the brickmakers, who seemed to feel it was in some way a rival for the fiery monster they'd created. Meanwhile, it transpired that despite their many reassurances they did not have enough firewood for the kiln. In fact, they only had about half what was needed. So one of them stripped the back seat from his ancient American saloon and drove off in search of more. Eventually, late in the evening, more wood arrived, though not really enough. Too tired to do anything about it, we recalculated to make allowances. Then we left Cesar in charge of the kiln and drove back to Piura.

'Please fire the kiln more slowly and don't touch my little field oven,' were my final entreaties.

In the morning it turned out they'd burned all the wood by five in the morning. And someone had opened my oven: in fact Cesar was using one of my sample bricks as a seat. Grumpily, I weighed my bricks anyway.

With nothing to be done on site until the kiln cooled down, we went with the brickmakers to the community centre. There we conducted a training session on the basic principles for energy-efficient brick firing and answered their queries. We explained briefly the differences between a kiln that had fuel in layers and one that was fired from the bottom, and gave some rules of thumb for coal replacement and distribution.

Finally, we made some calculations on comparative cost. Unfortunately, this showed that firing with coal was more expensive. This of course was not the full story, as wood is scarce and obtained illegally. Nevertheless, we needed to reconsider whether or not coal *was* an option for brickmakers in La Huaca, which was embarrassing. Overall, however, the session went well. The brickmakers were very interested and much friendlier once they understood something about what was going on. I wished fervently that we'd had this session *before* they'd nuked the kiln.

I complained sadly to a colleague that brickmakers did not tell me the *truth*: their facts and figures were contradictory. He explained patiently that a brickmaker was like a cook. Who knew exactly how much salt or chilli or garlic was used – a pinch here, a peck there? As long as the food was tasty, it did not have to be exactly the same taste every time. Like meals, bricks are not usually prepared in the laboratory.

8
THE PROJECT IN PERU AND ECUADOR – EXPERIMENTATION AND CONSOLIDATION

IN THIS CHAPTER, the course of the project in Peru is described, including the setbacks as well as the successes. The process of participatory working is examined in some detail. The principles of energy-efficient, cost-effective brickmaking that were evolved with brickmakers are then presented.

The specific Ecuadorian project experience is discussed. Here the project was working with the poorest small-scale brick producers and the account describes the steps that were necessary to gain the confidence of the brickmakers and work with their participation.

An inauspicious start – the Cajamarca coal-fired clamp

In 1986 ITDG fieldworkers spent time gathering technical and socio-economic information about brickmaking communities in the three of the four areas in Peru where the project planned to work: La Huaca, Cajamarca and Ayacucho. These areas had been selected based on the number of brickmakers operating, the different firing methods and fuels already being used, and the interest shown by producers in participating in the project. As Box 8.2 illustrates, time spent listening to what the brickmakers identify as their problems is always time well spent: it means that the brickmakers give the project more of their commitment because they perceive the project to be addressing *their* problems.

The first landmark of the project in Peru cannot, however, be described as a success. The brickmakers of Cajamarca had expressed an interest in experimenting with substituting coal for fuelwood. With the assistance of an engineer from ITDG in Zimbabwe, in 1986 the project team built a coal-fired clamp near the town of Cajamarca. The clamp did not burn. Although the team might have expected the new technology not to be an instant success and to require adaptation for the new context, this initial failure presented some immediate difficulties. The brickmakers who co-operated in building the trial clamp were not impressed. Despite lengthy discussions and explanations, first impressions are important: this was not a good face for the project to present. Furthermore, project resources – the money available – would not stretch to another visit by the experienced Zimbabwean engineer. And they would not cover too many of these 'failures' in research. The fact that the technology that was supposed to be the core of the project was judged – partly at least – a failure, and that fresh technologies had to be developed, undoubtedly changed the nature and timescale of the project.

Being underfunded is the seemingly inevitable reality for many development projects. Unlike their counterparts in some commercial companies, the project team could not simply throw money at the problem to solve it. What they could do was to talk openly with the brickmakers about the principles of the clamp and explain the suspected reasons for ignition failure. These frank exchanges, which took place during the discussion session that followed, actually served to cement the relationship between the project team and the brickmakers. The team had no wish to present themselves as infallible: they were always ready and willing to admit that they did not have all the answers. The outcome was that brickmakers felt more connected with the project and could see clearly that their involvement and commitment were needed if such problems were to be overcome and successful technologies developed.

Also on the positive side, it was at this point that it became crystal clear to the team that there was a need to find a way of monitoring the energy consumption of clamps and kilns to compare the efficiency of burning processes. In retrospect, identifying this need and ultimately resolving it was to be one of the major technical achievements of the project.

Gathering together Latin American experience

After the Cajamarca clamp, the project team extended contacts with brickmakers in other areas of the country. Preliminary work had concentrated on Cajamarca and Alto Mayo regions. Now the team visited brickmakers in La Huaca and Ayacucho, and began to form working relationships in both areas. They also visited brickworks of all scales in other parts of the country, studying carefully the commercially successful practices they encountered and noting the possibilities for small-scale brickmakers.

In an international workshop, held in Piura in early 1997, interested parties from Zimbabwe, Peru, Ecuador, Britain, Cuba, Brazil and Colombia came together to share experiences. The forum included engineers and brickmakers, social scientists and academics, environmentalists and representatives of training organizations. A variety of presentations were made and wide-ranging debate took place, both in the large forum and in topic groups. Alliances were made, relationships cemented, and an information and liaison network formed.

The next technical step was a pragmatic one. The team decided that the key to success was definitely attempting 'small changes'. History is full of stories of much-hyped new versions replacing well-adapted older versions, with disastrous consequences (see Box 8.1). Local practice should neither be discarded nor altered too quickly. Realizing that a way of measuring energy efficiency was needed, and beginning to evolve that methodology, the team became even more deeply aware that current practice might *be* the best practice. It would certainly have elements that were worth retaining or developing: there was a local knowledge to be learned from. Perhaps the issue was not bringing in an exogenous technology at all. Perhaps it was more a case of making the right technology available to the right people in the right place. Maybe there were brickmaking practices within Peru – or within Latin America – that simply needed to be observed, evaluated and disseminated.

Another factor that was thought to be vital to success was ensuring the whole project team – individuals from Peru, Ecuador, Zimbabwe and Britain – approached the project from the same standpoint. Engineers and technicians from the different countries had a wealth of knowledge about brickmaking and energy use but, thus far, had not had the opportunity to share and develop it practically. It was essential that the baseline of knowledge about brickmaking technology was common. Perhaps even more essential was establishing a

> **Box 8.1** An example of a development disaster – replacing the Haitian Creole pig
>
> The history of the eradication of the Haitian Creole pig population in the 1980s is a classic parable of globalization. Haiti's small, black, Creole pigs were at the heart of the peasant economy. An extremely hearty breed, well adapted to Haiti's climate and conditions, they ate readily available waste products, and could survive for three days without food. As much as 80–85 per cent of rural households raised pigs; they played a key role in maintaining the fertility of the soil and constituted the primary savings bank of the peasant population. Traditionally a pig was sold to pay for emergencies and special occasions (funerals, marriages, baptisms, illnesses and, critically, to pay school fees and buy books for the children when school opened each year in October).
>
> In 1982 international agencies assured Haiti's peasants their pigs were sick and had to be killed (so that the illness would not spread to countries to the North). Promises were made that better pigs would replace the sick pigs. With an efficiency not since seen among development projects, all of the Creole pigs were killed over a period of 13 months.
>
> Two years later the new, better pigs came from Iowa. They were so much better that they required clean drinking water (unavailable to 80 per cent of the Haitian population), imported feed (costing US$90 a year when the per capita income was about US$130) and special roofed pigpens. Haitian peasants quickly dubbed them 'four-footed princes'. Adding insult to injury, the meat did not taste as good. Needless to say, the repopulation programme was a complete failure. One observer of the process estimated that in monetary terms Haitian peasants lost US$600 million dollars. There was a 30 per cent drop in enrolment in rural schools, there was a dramatic decline in the protein consumption in rural Haiti, a devastating decapitalization of the peasant economy and an incalculable negative impact on Haiti's soil and agricultural productivity. The Haitian peasantry has not recovered to this day.
>
> *Source:* Aristide (2000)

common way of working with brickmakers. The team had to adopt the Ten Commandments of technology transfer and participation, and develop a unified approach. No one, for example, wanted to falsely raise the expectations of brickmakers.

The project in La Huaca

The next step was to bring together team members from Peru, Ecuador and Zimbabwe to work with brickmakers in La Huaca. The aims of the exercise were:

- to burn a trial kiln, using a mixture of fuelwood and coal in layers;
- to test the methodology of monitoring energy efficiency in practice;
- to exchange knowledge and strengthen links with the brickmakers; and
- to consolidate the team.

In May 1997 what became known as a 'hybrid kiln' was duly fired, using the Scotch kiln of Señor Victor Carmen (this event is described in more detail in Box 7.5). The fuel mix was different from the coal-fired clamp, and the bricks were being stacked inside a permanent kiln of the kind commonly used in this part of Peru rather than in a temporary clamp. Employing the principle of the Zimbabwean clamp, part of the fuelwood normally used to bottom-fire the kiln was replaced by coal placed between the layers of bricks. Brickmakers in La Huaca had a desperate problem getting firewood. The local area was deforested and it could only be obtained – illegally – from quite distant forests. A brickmaker from Ayacucho made the long trip north to invest the process with his knowledge. In Ayacucho there was a tradition of burning with coal. Thus, after extensive discussions, the assembled allies decided to try to replace 40 per cent of the energy normally obtained from firewood with energy from coal dust. (In fact, when all the calculations had been made, it was found that 37 per cent of the total energy came from coal.)

It was an intensive process: testing the rigorous methodology for the first time and firing in a new way with at least half a dozen different opinions about the best way to do it. In spite of the difficulties, the kiln burned well, producing bricks of a quality that compared favourably with the brickmakers' normal output. Meanwhile, the monitoring methodology worked; snags were ironed out and modifications made. The energy consumption of the kiln compared favourably with firing with 100 per cent fuelwood: the specific firing energy consumption of the hybrid kiln was 2.40 megaJoules per kilogram (MJ/kg) of fired brick, while that of a directly comparable kiln fired using only fuelwood was 3.40 MJ/kg. No conscious attempt had actually been made to reduce the overall energy consumption of the process. It was thought that the improved efficiency that should result from the distribution of coal as fuel throughout the clamp would be reflected in improved brick quality. To an extent, this also happened. It was noted, however, that to gain the full benefit of using coal in such a way, the techniques of kiln firing would need to be modified.

With bottom-fired kilns, the fire needs to be very hot if heat is to be transferred to the topmost bricks. Hence, small-scale brickmakers keep a raging fire going for a sustained period (though they typically do not hold the critical temperatures needed for vitrification for long enough). When fuel is distributed throughout the kiln the firing can be slower and less intense. Once the fuel in one layer is ignited it should heat the bricks above it until the next layer of fuel is ignited – and so on. In Victor Carmen's trial kiln, burning in the traditional way served to ignite all the layers virtually simultaneously and some advantage of the new method was lost.

In a workshop that followed the kiln firing, the assembled brickmakers and engineers evolved ten rules for energy-efficient, cost-effective brickmaking: these are listed at the end of this chapter. The assembly also discussed the different firing techniques required for kilns that had fuel in layers. Lastly, they compared the cost of firing the hybrid kiln with the normal wood-fired method.

From this comparative analysis emerged another seeming setback: despite the energy saving, fuel for the hybrid kiln had cost marginally more than using solely fuelwood. It was not the cost of coal *per se* that was a problem, it was the cost of transporting it to La Huaca. Although bulk delivery ought to reduce the cost, the prospect of ordering in bulk from distant merchants did not appeal to the brickmakers. These men – all the brickmakers at La Huaca were men – were not the poorest brickmakers. They were operating commercially, but on a limited, semi-formal scale. They had customers who came to buy bricks from them, but they did not have order books or regular outlets for their bricks. Working as individuals meant

> **Box 8.2** Listening to the people's concerns
>
> 'First, you have to get to know people and know the local situation. You do not come running in and say, this is what we are going to do. You learn about the conditions people face; you listen. For example, in La Huaca brickmakers told me about the situation with respect to fuelwood. When I had listened to what they had to say individually, I repeated it back to them all in a workshop to check. I said, "So you cannot cut trees for fuel because the police know who brickmakers are and they will stop and fine you?" Everyone agreed that this was the main problem. Permits are needed even to cut trees that stand on your own land.
>
> Then I asked, "Do you want to try something different, something that could give you an alternative to fuelwood?" The brickmakers agreed that they *had* to find an alternative. You have to base your intervention on people's concerns – plant it in people's reality. In La Huaca, 40 per cent of the people are involved in brickmaking – the community depends on it. So, anyway, after this we were united in looking for fuel options and testing them: the brickmakers knew I was there to assist them; I was not going to do anything that they did not agree with.
>
> It is also important to gain people's confidence. The initial period of getting to know people and understanding their socio-cultural and economic reality is vital. If possible, I like to fit in with existing social structures. In La Huaca, the local municipality paid for exchange visits, and the priest donated roofing sheets so that the brickmakers could build a shelter to house their equipment. The priest has often helped us to understand the local situation. He has been here for many years; people have confidence in the priest.'
>
> Jorge Marquina, ITDG field engineer

that they were hampered in seeking to supply, for example, contractors engaged in larger building projects. In short, the brickmakers would find it difficult at the stage their businesses were at to start accounts with coal merchants and to pay in advance for deliveries. The knowledge that obtaining wood exposed them to the danger of heavy fines and that they were aware of the on-going environmental damage which was taking place due to deforestation did, however, mean that they were anxious to find alternative fuels. But in 1997 they were not yet ready to use coal, with all the organizational changes that it would involve.

Further innovations: sawdust and rice husk

It is worth following the story of the La Huaca brickmakers through to its conclusion – at least in terms of the ITDG project, which ended in March 2000. The brickmakers in La Huaca were relative newcomers to the industry. They had been farmers but, due to a natural disaster, the area had become unsuitable for agriculture. Exhibiting a marked flexibility and resilience, a number of people became brickmakers. Being relatively new to brickmaking and having already demonstrated their ability to adapt, the La Huaca brickmaking community became some of the project's most committed, ingenious and progressive allies. Together the brickmakers and the project team sought an alternative to scarce fuelwood and expensive coal, and they found one – sawdust. The project had observed that brickmakers in Pascuales, near Guayaquil in Ecuador, incorporated a large percentage of sawdust into their bricks as the

88 BRICK BY BRICK

Photo 8.1 The rice husk in front of this Scotch kiln is used as a cheap fuel source

Lucky Lowe/ITDG

principal fuel. Sawdust was available from sawmills at a price that made it competitive in La Huaca. Drawing on the project's international experience, brickmakers at La Huaca quickly adapted to mixing sawdust into the soil they prepared for brickmaking (see Box 8.3).

This innovation took off so well, that by early 2000, using sawdust in this way was common practice with all the brickmakers in the community. Some wood was used to ignite kilns but the main fuel was sawdust. Brickmakers at La Huaca and the project team continued to experiment with the use of agricultural residues – particularly rice husk – as a substitute or partial substitute for sawdust. Rice husk is a waste product that the brickmakers in La Huaca had themselves tried adding to bricks before the project began: it is even more readily available than sawdust, and cheaper. What the project was able to provide was the methodology to monitor the energy used in the various rice husk and clay mixtures the brickmakers were trying out.

The community in La Huaca is determined to improve both the quality of their products and the sophistication of their business practices. They want to address larger, more lucrative markets; and they want to supply their customers directly – at present they lose out to transport and retailing middlemen. So, brickmakers want their own transport to bring in fuel and to deliver bricks and roof tiles. They want to bring their production up to the quality required for government and private sector building projects; and they want to work out a system of collective marketing so that they can tender for large supply contracts. They have set up their own association, the Yawar Huaca Brickmakers Association, and it is this organization that must play an important role in these collective activities (see Box 8.4). It takes time for an organization of this kind to become strong enough to take on these tasks, however. In the meantime, individual brickmakers are increasing their knowledge and developing their skills; they are improving their product, and tailoring it to the market. They are innovative and dynamic, hard-working and ambitious. They are a community that thoroughly deserves to succeed and prosper.

More fuel alternatives: coal briquettes and waste engine oil

Elsewhere, ITDG staff began to experiment with the low-pressure 'briquetting' of coal dust. In this process, coal dust is formed into balls – a small percentage of clay being used to bind the ball together. This fuel was seen as an option by brickmakers in some areas, particularly Ayacucho. Brickmakers there carried out many trials in alliance with ITDG and monitored the energy efficiency of the kilns.

Meanwhile, ITDG staff in Peru discovered a technology that exhibited great promise for brickmakers – not only in Peru, but also around the world – for whom wood, coal, sawdust, briquettes and even agricultural residues were not viable fuel choices. In the city of Trujillo, which lies north of Lima in the coastal region, ITDG staff met Señor Luis Alvarez Rodriguez, who was to become one of the project's most crucial allies. Luis Alvarez is a contract oil-burner: he fires kilns for brickmakers.

The technology Luis Alvarez employs is relatively simple. He burns waste engine oil, which is available in most urban centres. The oil is gravity fed through pipes and tubes to the bottom of the kiln. A compressor pumps air through a parallel system of pipes, the oil and the air are emitted through jets, the mixture is lit and the flow of air adjusted to obtain efficient combustion. The oil-burner is made so that the fuel-air mixture is jetted into the firing tunnels that are traditionally built into Scotch kilns. This is used in the following design:

Box 8.3 Trying out sawdust

'Books are okay as a guide, but experience is the only real teacher. There is always a difference between theory and practice. Soils do not perform as theory dictates. Here we found that when we added sawdust to the soil, the bricks we made shrunk more when they were drying. Why is that? The books say that adding sawdust will reduce the drying shrinkage. You have to make allowances; you have to learn. We had to adjust the size of the brick moulds so that the finished – fired – bricks came out at the standard size that the market requires. In our work here, we have built on local knowledge. For example, we came to try sawdust as a fuel because Francisco – one of the brickmakers in La Huaca – remembered seeing it used somewhere else in Peru. I told them that ITDG had some experience and that we were working with brickmakers in Ecuador who used sawdust. Then I said to the group, "So shall we try it here?" They said, "Yes"; and so we did.

There were problems of course. The sawdust we get varies – it comes from different sources. Often it contains shavings that have to be picked or sieved out. They can cause cracks or imperfections when bricks are drying or being fired. Some sawdust is from indigenous hardwood trees and some is from softwoods. The amount that is mixed into the brickmaking soil has to be adjusted so that bricks are well-burned and of good quality. I can use a laboratory to test the calorific value and moisture content of sawdust, but this is too expensive and time-consuming for brickmakers to do every time they fire a kiln. So we combine my scientific knowledge with experience. We learn that sawdust from this or that sawmill, that comes from this or that wood, and is perhaps very wet, should be used in a certain proportion.

After the initial period, it is results that breed confidence. It's difficult doing research work in the field. Brickmakers are looking for improved quality or cost savings – or both! As researchers we may be happy with a failure – it may prove a point – but brickmakers find this difficult to understand. So, we have to gear our fieldwork with them towards success. And we have to be prepared to compensate for failure. The brickmakers can't afford to lose 30 000 bricks because I've recommended trying to burn with rice husk as a fuel and then it hasn't worked for some reason.

You always have to follow up too; brickmakers need to know when you'll be around and how to contact you. For example, one of the reasons that coal was not adopted as a fuel at La Huaca was because we were not around at the crucial time to advise. Coal had worked quite well in initial trials, and at that time it was the only alternative to the risk of stealing fuelwood that brickmakers had. But we were not around, so brickmakers bought the wrong sort of coal and kilns did not burn well. This discredited coal in the eyes of the brickmakers – even though it could have been okay – simply because no one was here to offer advice at the right time. It was logical for them to choose the cheapest coal. With the information they had, I would have done the same.'

Jorge Marquina, ITDG field engineer

> **Box 8.4 Lessons from working in La Huaca**
>
> One of ITDG's field engineers in Peru provides some insights into the everyday reality of working with brickmakers.
>
> 'What are the essential elements of a process of technology development like this? If I had to list them, I would say:
>
> - Intervention has to be firmly based on the reality of local needs.
> - You cannot ignore personalities or personal agendas either – those of the brickmakers or your own!
> - You have to choose the right person to work with to achieve specific goals.
> - Appropriate, practical training is vital.
> - Participation is essential – working together to a plan that has been discussed and agreed.
> - A good project manager must maintain an overview, communicate that to fieldworkers and brickmakers, and motivate them.
> - We also have to be flexible: the brickmakers may want to change something, and we have to work with that and see it as an opportunity.
> - Possibly the most important thing is institution building to strengthen the Brickmakers' Association. Joint marketing would improve the lot of the brickmakers even more than better technology. They need to supply the market with bricks of the right quality, delivered on time; they need to regulate their prices and not undercut each other; they need to do something about the cost of transport, which is killing them. We know this requires institution building, but this takes time. I can advise, but the brickmakers will do this for themselves when they are ready.'
>
> Jorge Marquina, ITDG field engineer

- bricks are loaded into the kiln in strata interspersed with layers of coal dust;
- charcoal briquettes (made of clay and coal dust) are placed in the kiln's firing channels, throughout the kiln;
- the oil-fired burner is used to ignite the briquettes.

Despite being quite common in Trujillo, the oil-burner technology was unknown in the other areas of Peru where ITDG worked with brickmakers. Enlisting the support of Luis Alvarez, ITDG arranged firings with the oil-burner in Cajamarca, Ayacucho and La Huaca. It worked well: brickmakers were impressed and interested; Luis Alvarez proved an able demonstrator and a good trainer. The oil-burner was copied in La Huaca and improved upon, others were produced and an extensive series of monitored kiln-firing trials embarked upon. Eventually five oil-fired engines with air pumps and oil jets were constructed and are in use in La Huaca, Ayacucho, Cajamarca, Chambo (Ecuador) and Tunia (Colombia); there is also evidence that the burners can be repaired and maintained. This suggests that the technological capability to use the oil-burner is established.

Luis Alvarez is quick to say that in the process of disseminating his knowledge he also learned a great deal. He has seen brickmaking technologies he did not know existed; and he has met and talked with brickmakers and technologists from all over the world. The

Photo 8.2 Making coal-dust briquettes to substitute for fuelwood

Emilio Mayorga/ITDG

Photo 8.3   Learning by doing: visiting brickmakers observe the use of the oil-burner for igniting a kiln

Emilio Mayorga/ITDG

interaction has given him many new ideas and he has made many constructive suggestions about all aspects of the project. Before he became involved with the project, Luis Alvarez had not travelled far beyond Trujillo.

The time taken to ignite the kiln using an oil-burner is two hours: this compares favourably with the use of fuelwood for ignition, which takes 12 hours. This technology is also cheaper than using fuelwood: costs were reduced by 50 per cent.

The use of coal briquettes together with an oil-burner was comparable in cost with using sawdust mixed into the body of the bricks. The principal advantage of briquettes over incorporated sawdust, however, was the improved quality of the product: the bricks fired in this way were found to have twice the compressive strength of those obtained with sawdust. On a cautionary note, such results, obtained in a limited number of cases, are not rigorously comparable. Thus, it certainly must not be assumed that firing with briquettes is going to be better than using sawdust in all situations.

In some cases, where fuelwood was bought illegally and was increasingly unobtainable, energy efficiency and even cost-effectiveness become irrelevant. Brickmakers have no choice but to use alternative fuels. In the latter stages of the project during 1999, kilns that utilized a combination of the oil-burner, briquettes and coal dust used in 'cool spots' – the corners – were tested. Results compared favourably in terms of energy efficiency with other practices.

Developing hardware – crushers, extruders and oil-burning engines

The processing of clay – extraction, mixing and moulding – is performed manually by almost all small-scale brickmakers in Peru and Ecuador. The work is arduous and brickmakers generally exhibit a keen interest in mechanized extraction, mixing and moulding or extruding. Whether they have sufficient capital to invest in machinery and whether the rate of return on that investment can be justified remains an open question. Will mechanization increase output, improve product quality and hence the price obtainable? Will it increase income? And will it mean more jobs are created? It is commonplace enough for technology solutions to be pleasing in the abstract, but to benefit only better-off producers and to result in jobs being lost (see Box 8.5). These are all questions that must be borne in mind, even if they cannot be answered with any certainty before trying out the equipment.

It is difficult to convince brickmakers that improving product quality, extending their product range and being able to address more lucrative markets can be achieved without investing in hardware. Perhaps in common with small-scale producers everywhere, machinery is identified with progress and an associated increase in status. In response to the expressed needs of the brickmakers, therefore, the project diversified its energy efficiency activities somewhat. The brickmakers wanted to test various equipment options. After lengthy and careful discussion, they decided that their priorities were a manual clay crusher, a manual extruder and a small motor-driven extruder. All the brickmakers acknowledged the need for training to accompany the introduction of new processes and hardware.

The purchase of equipment for testing was duly arranged. Manual and motor-driven extruders were sourced in Peru, while the manual clay crusher – a proven design – was imported from Britain in early 1998. After trials, the brickmakers and ITDG were pleased with the clay crusher and the motor-driven extruder. The manual extruder, though it worked well enough, proved too slow to be useful to brickmakers working on a mass-production, commercial scale. In fact, the project team is of the opinion that, generally, manually operated machines for functions such as clay extraction, mixing and extruding do not perform well. Such tasks are better carried out either manually or by powered machines. Using animal power for clay mixing can also be a viable option.

The extent to which brickmakers adopt clay-processing machinery after the project team has left remains to be seen. The uptake of the oil-burner for igniting clamps was more assured: five were purchased in the main project areas, and the process of technology transfer was observable as it went through a number of stages. In each area:

- brickmakers observed the oil-burner being used during an organized visit to another area;
- experienced brickmakers were invited back to the area to demonstrate firing a clamp with the oil-burner;
- brickmakers purchased their own oil-burner and operated it themselves – often with difficulties that required assistance from experienced oil-burners;
- local brickmakers became confident at using and maintaining the oil-burner, and were able to advise brickmakers from other areas.

Training through exchange visits

Whether it was in the use of the oil-burner or in substituting fuels, organized exchange visits were key to the success of the technology transfer process. Practical experimentation at each of the three sites in Peru (plus two in Ecuador and two in Colombia) was important for local brickmakers to take part in, but dissemination also occurred as exchanges took place between the stakeholders engaged in work at each demonstration site, i.e. brick producers, firers, mechanics and equipment producers, technicians, professionals and their organizations. During the course of the project, from 1986 to 2000, exchange visits included:

> **Box 8.5 Beware the indirect cost**
>
> In Nepal, an ITDG pilot study was carried out for a new electric cooker, a heat-storage device. Cheap-rate night-time electricity or electricity from a small hydro-electric scheme could be used to 'charge' the cooker. The energy was then released as heat for cooking the next day. (Some types of electricity-generating installations cannot economically be shut down when there is no consumer demand. Hence, there is 'free' power available at certain times.) The cooker had quite a high investment cost but worked very well in technical terms. Heat storage made good use of surplus power and would be economic for people who paid a single flat charge for their connection to the supply.
>
> Being quite a sophisticated piece of equipment, however, only the better-off households could afford it. Their demand for fuelwood declined and there was a marginal benefit to the local forests. The fuelwood collectors, though, some of the poorest people in the community, lost business. Fortunately, the pilot study picked up this information, and it shaped the evaluation of the technology. This is a classic development dilemma: the economic use of power improves, fuelwood is saved, but fuelwood collectors lose income. In other areas of Nepal, cheaper and simpler electric cookers achieved a large take-up. These cookers used peak-rate, expensive power, but the investment was affordable. Where do technologists and energy planners go from there?

- two visits by La Huaca brickmakers to Ayacucho;
- one visit by La Huaca brickmakers to Trujillo;
- one visit by Trujillo brickmakers to Ayacucho;
- one visit by Trujillo brickmakers to Cajamarca;
- three visits by Trujillo brickmakers to Piura;
- two visits by Trujillo brickmakers to Chambo, Ecuador;
- one visit to Piura from Cajamarca, Ayacucho, Chambo.

It is to be hoped that dissemination of the more successful technologies will continue after the lifetime of the project through visits by brickmakers to different areas. Of course, the brickmakers themselves are quite capable of organizing for someone to come and provide technical assistance: for example, in overcoming a problem with the oil-burner. It is less likely, however, that brickmakers from a new area would become aware of the advantages of a new technology and organize themselves to visit and to be visited by experienced brickmakers.

For this to happen, the stimulus of demonstrating new technologies is ideally provided by local technical assistance organizations. It became clear during the course of the project, however, that institutions which might have had the potential to train brickmakers in new firing technologies after the end of the project were not wholly suitable. One Peruvian organization identified early on was more concerned with large-scale brickmaking; SENCICO in Peru and ITACAB in the region continued to work with the project, but were not constituted to deliver the sort of practical training brickmakers require at their place of work (see Box 8.6). In fact, there did not appear to be any such organizations on the ground in Peru.

> **Box 8.6** Appropriate training
>
> 'When it comes to training, there are things that work well and things that do not. Brickmakers are practical people. They do not like to read big books about the theory of brickmaking. On the other hand, they read the instruction booklet on how to use the oil-burner. That booklet is very well read because it is directly relevant. But you cannot write down everything. Dissemination works best with exchange visits and on-site training. It is about listening and talking, repeating things and learning. When I pose questions I give people time to consider and to answer. If it's a group, I let them debate the issue among themselves; I do not make an input unless I'm asked. Brickmakers like videos too; they like to see what people are doing in other countries. Unfortunately, there is only one VCR in La Huaca, and the tapes tend to get lost.
>
> The best way of training on-site – for example, with Luis Alvarez the oil-burner – was to bring him here to fire a kiln. Then he has to come back if necessary to follow up. When brickmakers try to use the oil-burner on their own they often encounter problems – of course they do, the technology is new to them. So they need Luis Alvarez to come back. Usually he has seen the problems before. He is very experienced and the brickmakers respect him. If he has not seen a specific problem before, he can usually sort it out. And I'm around to offer advice if I can. Training is about the people you are dealing with and their capabilities. How long it takes and what it involves depends on them.'
>
> Jorge Marquina, ITDG field engineer

The Ecuadorian perspective

In 1997, ITDG engaged with an institutional project ally, ESPOL (Escuela Superior Politecnica del Litoral, Guayaquil), in Ecuador. The main contact person was Dr Alfredo Barriga, who was first introduced to the project in 1996 and who presented a paper at the project's first international workshop on small-scale brickmaking in February 1997. ESPOL subsequently extended the alliance, recruiting colleagues from ESPOCH (Escuela Superior Politecnica de Chimborazo) to undertake some of the work.

The project in Ecuador concentrated on two particular geographical areas: Pascuales, Guayas Province, and Chambo, Chimborazo Province. In Pascuales, which is near the city of Guayaquil in the coastal region, energy for artisanal brickmaking comes partly from sawdust, which is mixed with brick clay in varying proportions. The sawdust not only provides thermal energy, but also works as a 'conditioner', making the soil easier to mould and reducing drying shrinkage. The remainder of the energy comes from fuelwood burned in the bottom of clamps.

Brickmaking in Pascuales uses a lot of energy. Although incorporation of fuel into the bricks should give improved efficiency, the kilns are very small, do not have fixed walls to provide thermal insulation and hence do not perform well. The firing time is one day or less, and clamps have a specific energy consumption that is around 2.70 MJ per kilogram of fired brick.

In addition to their small size, the low efficiency of clamps in Pascuales is probably linked to the inappropriate quality of clay to be found there. The clay has an organic matter content. This not only makes it not very suitable for brickmaking – fired bricks are often weak and friable – but also means the soil might be more suitable for agriculture. Brickmakers are, however, temporarily ensconced on waste ground that belongs to the municipality. They have no land of their own, so they have no choice but to try to exploit this soil for brickmaking.

The people who work in Pascuales come from poor, urban fringe areas. They have very low incomes. In general, they have had little education and brickmaking is perhaps their only possible source of employment and income. Brickmaking is carried out by family groups or by brickmakers working together in informal, convenient and often temporary alliances. There is no distinction as such between owners and workers.

In contrast, brickworks at Chambo in Chimborazo Province in the sierra, or mountain, region exhibit remarkably low energy consumption, estimated as one of the lowest in South America. Chambo kilns have fuelwood built into them between the layers of bricks to facilitate even burning. Fuelwood is placed in the corners in upper layers, which had been observed to be cold spots where bricks were under-fired. This distribution of fuelwood is more efficient than employing bottom-firing alone. Firing time varies from two to three days. Kilns have a specific energy consumption of around 1.50 MJ per kilogram of fired brick. The permanent structure of the Scotch kilns used gives good thermal insulation, which boosts efficiency. The fact that Chambo is at high altitude does not seem to impact markedly on burning efficiency.

Pascuales, Ecuador – finding the right entry point

It is always important to plan an entry point carefully, to make sure that a community's first impression is not of a project team that is intimidating, disrespectful or raising the wrong expectations. This was particularly important in Pascuales, Guayas, where brickmakers were very poor and without much education. Alfredo Barriga's account of events in mid-1997 (Box 8.7) demonstrates the importance of making the entry point low key, and of gathering information in an unobtrusive way. The order of events was as follows:

- A few team members visit the brickmaking community to introduce themselves, and to explain the purpose of the project.
- A week later, at an agreed time, the project team returns. During this meeting the brickmakers' understanding and expectations of the project are discussed, and the project is explained in more detail, being very precise about the team's motives and purpose. The brickmakers are encouraged to identify their problems.
- At the next meeting, data are gathered on the socio-economic position of the brickmakers and the technical practice employed. Questionnaires may be too intrusive (see Box 8.7) and techniques such as semi-structured interviews may be more appropriate. Choosing the right language, and making sure that people understand the same things by the same words is also important (see Box 8.8) At this point a working agreement is written down, detailing the roles and responsibilities of all participants of the project.

> **Box 8.7** The advisability of not rushing in
>
> 'With the brickmakers at Pascuales we first visited them simply to introduce ourselves. We did not attempt to do any technical work on that first visit. All our visits are low profile. We never arrive in a convoy of new vehicles with a team of social scientists and engineers, clutching clipboards and measuring equipment. We try not to bring too many visitors to the brickmakers: we do not want to overwhelm them. On that first visit we briefly set out the project in a clear and simple manner. Then we went away and let them have time to think about it.
>
> On the second visit, we looked for common ground. Obviously, cost saving – or rather increasing income – is of major interest to brickmakers. Increasing energy efficiency means saving fuel, which saves money and increases the profit brickmakers can make. So, here is common ground. But we did not say that brickmakers should increase their energy efficiency because deforestation is a global problem and we wanted to reduce emissions that were damaging the environment. That's not the language to use if you want to find common ground. A good exchange took place and we arranged another visit to gather some preliminary data and to have further discussions.
>
> We are very careful not to raise expectations. The brickmakers at Pascuales are very poor people. We arrive in a vehicle – even if it's not a smart new vehicle – and we are middle-class academics from the university who live in "better" areas of the city. It is obvious the brickmakers will expect something from us. And they should get it. But we repeatedly had to explain that we had no money to hand out, that we could not make grants or loans.
>
> Brickmakers have to have the time and opportunity to express their points of view and their concerns. Not everyone speaks up in a group. And it is not always the self-appointed spokesmen and the loudest talkers who have the best points to make. We always have conversational exchanges. We do not like to use questionnaires: the questions are often too blunt, and people become shy when they are confronted by clipboards and papers.'
>
> Dr Alfredo Barriga, ESPOL, Ecuador

Institution building – brickmakers' associations

The problems that the brickmakers face are as much organizational as technical. By the end of the project, widespread use of improved technologies in La Huaca, Peru, for example, resulted in fuel savings of around 30 per cent, and overall savings of 15 per cent (Lowe, 2000). These savings were likely to be swallowed up, however, when a reduction in the public sector building budget reduced demand for bricks. Local brickmakers began to reduce the price of their bricks in order to secure a buyer.

Brickmakers may be able to withstand the pressure to undercut each other if the associations they belong to are strong enough. The Yawar Huaca Brickmakers' Association was formed in La Huaca in 1998, and was registered in 1999, and the Pascuales Association is also seeking legal recognition (see Box 8.9). Apart from having the potential to control prices, such associations may be able to arrange collective supplies of coal, to market the bricks

> **Box 8.8** Using the right language
>
> 'We did not use patronizing, pretentious or political language. We do not talk about net specific energy or calorific values, participatory technology development, stakeholders or adding value. We do not use jargon. If there is a *need* to use some special, scientific language, then we spend some time explaining it to the brickmakers and listening to their queries. The thing is, the language is different, but much of the knowledge is similar. Communication is a time-consuming process; it requires patience on both sides.
>
> When brickmakers were defining the problems they faced, the terms they used did not always mean the same to them as they did to us. For example, they said they had a problem with brick quality. Now we might have assumed that by quality they meant compressive strength or water absorption – but no. They measured quality in terms of the look of the brick – its colour and its surface finish. Their most important definition of quality, however, was transportability. They suffered a lot of breakages in transport and so lost a lot of money. We listened to the brickmakers list their technical problems, but we did not offer advice until we had fully understood their meaning. And even then, not until they asked for it.
>
> Communication problems need not be deadly serious, though. Some brickmakers use a bullock to mix the soil for their bricks – the animal walks in a circle, treading the mixture. Once I asked a woman brickmaker if she used a bullock for this purpose – only the word for bullock is the same as the word for beast. So my question was: "Do you use a beast to mix your soil?" "Yes", the woman answered, "my husband."'
>
> Dr Alfredo Barriga, ESPOL, Ecuador

jointly and thus save on transportation costs, or to arrange exchange visits to demonstrate technical innovations. The Yawa Huaca Association has already been able to access bank loans from the Bancos Materiales for brickmakers to rebuild kilns in the aftermath of the El Niño flooding.

However, both Jorge Marquina in Peru and Alfredo Barriga in Ecuador point out that this organizational strength is not achieved overnight (see Box 8.4). Association members need training in small business management, in how to form and run an association, and in collective working. This was not achievable within the time span and remit of the project.

Ten rules for energy-efficient, cost-effective brickmaking

As well as substituting alternative fuels for wood, brickmakers attending a workshop in La Huaca in 1997 were able to come up with their own general principles for designing energy-efficient clamps (Mason, 2000).

(1) Bigger kilns or clamps are more efficient

The bigger a kiln or clamp is, the smaller is its surface area compared to its volume, and hence to the number of bricks in it. Heat is lost from the surface area of a kiln or clamp. So, if this

> **Box 8.9 Keeping a clear focus, but staying flexible**
>
> 'We always make sure our staff, allies and any visitors we take to the brickmakers are well briefed. We do not want brickmakers getting different stories and becoming confused. For example, when we took a team of filmmakers to Pascuales to make a video of the work, we told the brickmakers in advance and took considerable time to brief the filmmakers about the project and the situation of the brickmakers. Our whole team had to have a common agenda: they had to ask questions that had a focus in terms of the project; brickmakers do not have time for idle questions, they have a living to make.
>
> Nevertheless, if a project is to be successful then the agendas of both – all – the allies have to be addressed. Sometimes this has to mean changing your own agenda to accommodate the brickmakers' most pressing concerns. In Pascuales, for example, the brickmakers are keen to register with the municipality so that their enterprise can be recognized. Then they can get secure land tenure and have access to business loans. We had to help address this need, despite the fact that it was not part of our own brief. We had to do what we could.'
>
> Dr Alfredo Barriga, ESPOL, Ecuador

cooling area becomes proportionally less compared to the volume, then more bricks are fired for less heat loss.

Table 8.1 demonstrates how the ratio of surface area to volume decreases as kilns or clamps get bigger and hence, proportionally, how much less energy is needed. The table is for a cubic clamp built with bricks of 230mm × 110mm × 70 mm. To make calculations easier, air gaps between bricks are included within these dimensions. The clamps are considered as having four cooling faces. The heat loss from the top – in exhaust gases – and the small loss to the ground can be considered separately.

Table 8.1 Relative heat loss through the sides of different sized kilns

Length of side (m)	Cooling area, A (m^2)	Volume, V (m^3)	No. of bricks	Ratio A/V
2.62	27.46	18	10 000	1.53
3.30	43.56	36	20 000	1.21
4.16	69.22	72	40 000	0.96
5.24	109.83	144	80 000	0.76

Example

Señora Jara, a Peruvian brickmaker, was using 0.40 tonnes of coal to fire 1000 bricks. She fired 80 000 bricks per month in clamps of 10 000. The coal had a calorific value of 25 000 MJ per tonne, and a tonne cost of US$400. She estimated her process was 50 per cent efficient.

Photo 8.4 Through their brickmakers' association, producers can organize joint marketing and other collective issues

That is, half the energy actually fired the bricks and the rest was lost to the atmosphere. This is a typical efficiency for small-scale brickmaking. Of the 50 per cent of energy lost, Señora Jara believed that at least one-third was from the sides of the clamps. How much money could she save by firing a whole month's production in a single large clamp?

Burning 0.40 tonnes of coal means $0.40 \times 25\,000 = 10\,000$ MJ per 1000 bricks. So eight clamps of 10 000 bricks uses 800 000 MJ. Half of this, 400 000 MJ, is lost. A third of this, 133 333 MJ (17 per cent of the total), is lost from the walls at a cost of US$2133 (133 333 MJ ÷ 25 000 MJ/tonne × 400 US$/tonne). Referring to the table, the surface area of a 10 000 brick clamp is 27.46 m^2. Eight clamps means a total cooling area of 219.68 m^2, proportional to the US$2133 loss. The surface area of an 80 000 clamp is 109.83 m^2, around half. What does this mean in practice? Well, almost 17 per cent of the energy is lost from clamp walls, corresponding to 5.44 tonnes of coal per 80 000 bricks. If this loss is halved, the coal used can be reduced to 366 kg per 1000 bricks, saving around US$1 087 per month.

(2) 'Square' kilns or clamps are more efficient than 'rectangular' ones

A kiln or clamp with equal sides has a smaller cooling area than a 'rectangular' one of the same volume. So, efficiency is better because of reduced heat losses. From the table, a cubic construct of 20 000 bricks has a cooling surface area of $3.30 \times 3.30 \times 4 = 43.56$ m^2. A rectangular construct with the same number of bricks – the same volume – could be built by making it four times longer and halving the height and width. That is, it would measure 13.20 m × 1.65 m × 1.65 m. The surface area, however, would be $((13.20 \times 1.65 \times 2) + (1.65 \times 1.65 \times 2)) = 49$ m^2. So, in the rectangular structure the heat lost from the side walls could be 49/43.56 = 1.125 times more.

(3) Increasing insulation reduces heat losses

Heat is lost through the top, walls and, to a lesser degree, bottom of the kiln or clamp. Anything that reduces this loss increases efficiency. Thicker scoving (plastering with mud) on clamps will reduce losses, as will using fired or part-fired bricks in the outer layers. For kilns, thicker walls or a wall with an air gap will help.

(4) Placing fuel as close to the bricks as possible is most efficient

Obviously, if a brick is a long way from the heat, it will not 'burn'. Having the fuel closer to the bricks is more efficient. So, brickmakers can try distributing their fuel throughout the kiln or clamp, in between layers of bricks, rather than, say, burning it all at the bottom in tunnels. Placing some or all of the fuel *in* the clay mix can be very efficient. Incorporating, for example, coal dust or sawdust into the body of bricks is an established technique. In some brickworks, all the fuel needed is inside the bricks. The fuel chosen should be fine so as not to result in large voids in the bricks. But how much fuel can be moulded into bricks? This varies with the clay type, the fuel type the burning process. Some experts suggest 5 per cent of fuel by weight as a maximum.

Example

Señora Jara can buy coal dust of the same calorific value as her coal. The dust is cheaper because it is regarded as waste. So if she can use it she will save money. Using bigger clamps, she needs 366 kg of coal per 1000 bricks. Her bricks weigh about 3 kg, so this fuel requirement corresponds to around 12 per cent of the mass. That's probably too much to incorporate all of it into the bricks, but she could try using up to half her fuel as dust in the bricks and distributing the rest as coal throughout the clamp.

(5) Continuous kilns are more efficient than batch kilns or clamps

Continuous kilns typically use 'waste' heat to pre-heat green bricks. This means less heat is lost and firing is more efficient. Also, the structure of the kiln does not need to be heated up for each batch of bricks. Continuous kilns, such as a Hoffman Kiln or a Bull's Trench, are usually only feasible for brickworks making 10 000 or more bricks per day. There is a lack of simple, low-cost designs for smaller continuous kilns. Anything brickmakers do to use waste heat can, however, cut costs.

(6) Green bricks should be dry going into the kiln

If green bricks still contain a lot of water when they are placed in the kiln, then energy, and hence money, is wasted just to dry them. In hot, dry climates, bricks should be slowly but completely dried before firing by using free energy from the sun.

Example

Señora Jara used to rush her freshly moulded bricks to be fired. The bricks weighed 3.60 kg going to the clamp and 0.60 kg was just water. It takes at least 2.59 MJ to raise each kilogram of water to boiling point and evaporate it. So, Señora Jara used 15 540 MJ in each 10 000 brick clamp just to dry her bricks. This corresponded to 0.62 tonnes of coal and cost her more than US$248 per clamp. So, these days she sun-dries her bricks and saves that money.

(7) Fuel should be dry

As with the bricks themselves, if the fuel contains water, then energy is wasted to evaporate it. If using fuelwood, it should be dry and seasoned. Large, dense, slow-burning logs are generally better than the same mass of small, green twigs. Fuel should obviously be stored in a dry place.

(8) Good burning control saves energy and money

The temperature in kilns or clamps should rise quite slowly and constantly, otherwise heat is wasted. Also, if the temperature rises too fast, bricks may be damaged. The temperature should not be allowed to fall until firing is complete. As a rough guide, a 40 000 brick clamp fired externally might be slowly heated over a period of two days until no more steam comes from the top. It should then be fired for around four days until the bricks at the top are getting red hot. The kiln is then sealed at the top and the fires maintained for another day. After this 'soaking' stage, the kiln is completely sealed and allowed to cool. It may take a week or two before it's cool enough to open. Smaller kilns will burn more quickly, larger ones more slowly. Typically, small-scale brickmakers burn bricks much more rapidly than this and do not reach the temperatures needed for the times required to ensure sufficient vitrification. This is a major reason why so many bricks are under-fired and weak.

The flow of air through a kiln should be controlled: too much will cool the bricks and waste energy; too little and the fuel will not burn completely. A significant loss from some kilns is due to too much cold air being drawn in. Using a permanently set hearth, or fire-grate, and dampers, allowing better control of the firing process, could offer substantial savings. Protecting the fires from cooling winds by using windbreaks will also save fuel and help the kiln to burn evenly throughout.

(9) Record keeping is important

Unless it is known how much fuel, or more precisely energy, a particular burning process uses, it will be impossible to know when improvements have been made and to answer the questions: has less energy been used? Were costs reduced? There is no need to employ a system as complex as the energy monitoring methodology every time a kiln is fired – this would be uneconomic for an individual brickmaker. Keeping basic records of the process and any changes in fuel use and brick quality will, however, provide vital information.

(10) Replacing primary fuel with free or cheap waste reduces costs

If 'waste' materials can be used instead of some part of the expensive primary fuel, this saves money. Agricultural residues that might be used to partially replace fuel include rice husks, sawdust, straw, maize cobs and animal dung. Industrial wastes such as coal dust, boiler waste or pulverized fly ash often retain a high calorific value and are cheap. Brickmakers can check what is available.

Summary

For small-scale brickmakers perhaps the easiest step to improve efficiency and save money is to make sure that the bricks are dry before putting them in the kiln. Then it is possible, without too great an investment, to work on kiln control: following a good firing regime and controlling airflow. Increasing insulation, particularly of clamps, which can simply be scoved with a thicker layer of mud, offers a potential saving without a big cash outlay. Substitution of wastes for part of the primary fuel and incorporating some fine fuel into the body of bricks can be done on a small, experimental scale to minimize the cost of any failures. Keeping good records only costs time and effort.

9
PROJECT OUTCOMES

THIS CHAPTER presents the results and findings of the project, concentrating on training, sustainability, technology, the environment and information dissemination.

The essential nature of the project changed during its course. The South–South transfer of coal-fired clamp technology from Zimbabwe to Peru and Ecuador was replaced by the development of local technologies, which were more relevant to the situation of small-scale brickmakers. Hence, work focused on evaluating energy efficiency, developing and disseminating technologies. Meanwhile, the process of participation, working with brickmakers and other allies, was effectively pursued. This process will be dealt with in more detail in the concluding section, 'Guidelines for participatory projects'.

This project managed to meet most of its targets despite a number of significant factors that impeded progress. The suspension of ITDG's Building Materials and Shelter Programme in Peru and the appointment of a new project manager, seconded from the Energy programme, resulted in some initial delays. The project proposal was – almost inevitably – over-ambitious. In addition, though, it was difficult to interpret and to implement. Ultimately, however, the project benefited from improved management and reinterpretation involving all stakeholders. The effects of the El Niño phenomena caused a good deal of work to be suspended or cancelled.

Training: getting it right

Despite these setbacks, only one of the project's principal predicted outputs – the establishment of a local Peruvian institutional capacity to implement a national training programme – was not achieved, and this was because it did not prove to be feasible. The educational and socio-economic strictures of most brickmakers in the 'target group' meant that the majority would not have the background, time or money to attend a formal training programme. Nevertheless, two Peruvian training organizations, ITACAB (Instituto de Transferencia de Tecnologias Apropiadas del Convenio Andres Bello) and SENCICO (Servicio Nacional de Capacitacion para la Industria de la Construccion), have been closely involved with the project throughout. Hence, such a programme has been discussed in depth with a view to adapting the concept for future implementation.

Brickmakers generally require training to be brought to them and delivered in a very practical way. Those defined as 'small-scale' brickmakers are not an amorphous group. They vary from peasant farmers with little formal education, producing bricks as a part-time activity, to entrepreneurs managing vibrant commercial concerns. Designing a training programme, to cover all aspects of brickmaking to be delivered to such a diverse audience in the field is a major challenge.

> **Box 9.1** The project's institutional allies
>
> The following local institutions collaborated with the project mainly by way of information sharing and promoting research findings:
>
> - Instituto de Tranferencia de Tecnologías Apropiadas del Convenio Andrés Bello – ITACAB / Perú
> - Servicio Nacional de Capacitación en la Industria de la Construcción – SENCICO / Perú
> - Escuela Superior Politécnica del Litoral – ESPOL / Ecuador
> - Escuela Superior Politécnica de Chimborazo – ESPOCH / Ecuador
> - Fundación Ecuatoriana de Transferencia de Tecnología – FEDETA / Ecuador
> - Servicio Ecuatoriano de Capacitación Profesional – SECAP / Ecuador
> - Corporación para el Desarrollo de Tunía – CORPOTUNIA / Colombia

Sustainability: what next?

Links have been established with relevant government ministries and the issues pertaining to training with the small-scale brickmaking sector are receiving increased attention. So, whether or not the training started in the project comes to a halt or continues in a sustainable manner remains to be seen. What is certain is that those brickmakers who have adopted either technologies that improve their energy efficiency or a fuel alternative to woodfuel are likely to continue doing so. Indeed, it is probable that some technologies, particularly the oil-burner and the use of sawdust, have gathered sufficient momentum and are sufficiently proven that they will continue to spread through an informal, 'copycat' process.

In human and organizational terms, links have been made between brickmakers in different areas of Peru and Ecuador, and between these brickmakers and institutions concerned with energy efficiency, brickmaking or training. These links are, however, only likely to continue in an informal way in the absence of financial resources and an organization to play the role of proactive facilitator that ITDG has been fulfilling.

Participants in the final project workshop in Piura expressed their interest in continuing the interaction. They were motivated to do so and proposed ideas for a phase two of the project. This proposal was firmly based on the expressed needs of the brickmakers who attended the workshops. Sustainability will ultimately depend on the mobilization of the actors who included, apart from brickmakers, entrepreneurs, NGOs, parastatal training institutions, universities and technical colleges. The project has certainly influenced these actors and raised levels of interest in improved, energy-efficient brickmaking and its benefits.

Monitoring technological change

Using the energy efficiency methodology, the project monitored 48 brick firings:

- 29 in Peru (19 in Piura, 9 in Ayacucho, and 1 in Cajamarca);
- 17 in Ecuador (14 in Chimborazo and 3 in Pascuales); and
- 2 in Cauca in Colombia.

Photo 9.1 Eventually the brickmakers of La Huaca could operate the oil-fired burner without external help

Lucky Lowe/ITDG

The main problem with compiling and analysing the data is that there are too many variables. Thus, it is difficult to get reasonable comparisons between the different firings in different places. Undertaking research in the field with real people involved in commercial concerns is not ideal for replicating trials: the brickmakers' livelihoods naturally take priority over research goals.

So, despite the number of firings completed, there were not sufficient results to begin a meaningful statistical analysis and to look into the degrees of correlation between the different parameters. For this reason the new technologies cannot be said to be fully proven. To prove, for example, that a particular technology gives a guaranteed improvement over the traditional technology would mean repeating many similar firings under carefully controlled conditions and doing so in a number of locations. What is apparent, however, is that some of the technologies developed *can* increase energy efficiency and potentially cut production costs – not only for small-scale brickmakers in Peru and Ecuador, but worldwide.

The dangers of not comparing like with like

Before noting three technologies that show great promise, some technical caveats must be made. As well as the fuel type and the kiln design, the soil type and its vitrification temperature appear to have a great bearing on the specific firing energy. Some clays may contain a small amount of carbon or other fuel element naturally, and so the external fuel required is reduced. Yellow Fletton bricks, found over much of Eastern England, for example, require only about one-third of the energy input of heavy clay bricks. It is impossible to compare the efficiency of firings where different clays were used and where the types of clays varied significantly and where intrinsic calorific values have not been recorded.

A significant number of bricks were under-fired in at least half of the firings reported. For future research, it could be recommended that where more than 20 per cent of bricks are under-fired or unusable, the results should be discarded. Otherwise, it would be all too easy to assume that these firings were highly efficient. The average figure for a theoretical 100 per cent efficient firing (one in which no heat is lost) might be taken to be 850 kJ/kg of fired brick (Barriga). A few of the project trial firings are around or even below this figure, but with large numbers of waste bricks. Where a kiln produces 90 per cent usable bricks, this may be an optimum proportion for comparing like with like.

On the other hand, firings with 95 per cent usable bricks seem to demand excessive fuel use and even, in a few cases, produce overburnt bricks – even with these relatively basic field kilns. In other words, to get that additional 5 per cent of bricks to be usable, a lot more fuel is needed. The bulk of bricks, meanwhile, are being heated to well above their vitrification temperature without markedly improving their properties.

It is neither possible nor desirable to present the project's results in full, but comparing three of the monitoring forms may be instructive. Monitoring forms 01 (Appendix 2) and 07 (Chapter 7) both refer to a similar-sized kiln with the same operator and utilizing the same clay, and the main difference between them is that for Form 01 coal dust has been substituted for part of the fuelwood and coal has been distributed between the layers of bricks rather than relying purely on bottom-firing. By doing this, a significant improvement in energy saving has been achieved. Even greater savings can be expected if the coal dust, or indeed sawdust, is incorporated into the body of the bricks.

Table 9.1 Specific firing energy employed using different fuels and different-sized kilns

Monitoring form no.	No. of bricks fired	Firing technology and type of fuel	Specific energy
07	6350	Traditional wood kiln	3.40 MJ/kg
01	6350	Wood, coal dust and coal distributed between the brick layers	2.4 MJ/kg
14	15 000	Kiln adapted for project with anthracitic coal dust distributed between the layers, and oil	1.58 MJ/kg

Meanwhile, monitoring form 14 (Appendix 2) shows the efficiency that can be obtained in a bigger kiln, employing coal distributed in layers and using the oil-burning technology developed by the project. Obviously, it also shows that artisanal kilns can be fired without the use of any fuelwood whatsoever, obtaining good efficiency and comparable brick quality.

Technologies: three to consider

Technologically, the project has made significant strides forward in working with the small-scale brick production sector. Developing a standard methodology for measuring and comparing the energy efficiency of brick-firing processes is an important achievement with international applicability that merits widespread advocacy and dissemination.

Incorporating residue fuels into the body of clay bricks is a particularly promising technological development that ITDG intends to pursue. A project funded by USAID researching the use of rice husks as a brickmaking fuel is underway in Peru. The brickmakers of La Huaca have already adapted to using sawdust incorporated into the body of bricks as their principal fuel. This technology is successful and likely to be adopted in other regions where sawdust is available.

The waste oil-burner is also worthy of further research and development. The oil-burner design could be made more flexible, easier to regulate and cheaper to manufacture. In areas where it is prohibited to use fuelwood for brickmaking, and coal and sawdust are unavailable, burning oil may be the only option open to brickmakers. Time will tell how many oil-burners are manufactured for and purchased by brickmakers. Alternatively, some entrepreneurs could become professional, contract kiln burners, like Luis Alvarez Rodriguez. In 2000, the oil-burner cost around US$3000 and only the better-off brickmakers or coalitions of brickmakers could afford the capital outlay. A contract burner could service a good number of brickmakers, allowing a rapid return on investment. Without a phase two of the project, which could consider facilitating the provision of credit to such brickmakers or entrepreneurs, the technology is likely to be sustainable only for the better-off.

Hand-moulded fuel briquettes. Commercially produced, high-pressure briquettes were already used by some commercial brickmakers, notably near the city of Trujillo. Hand-moulded briquettes, though, can offer smaller-scale brickmakers a solid fuel alternative to wood or coal. Caution is needed, however: making briquettes that burn well and which do not break up takes practice. Generally, hand-moulded, low-pressure briquettes are not suitable for distributing throughout a clamp or kiln because they crush quite easily; however, kilns that utilized a combination of the oil-burner, briquettes and coal dust used in 'cool spots' – the corners – have been tested in the latter stages of the project. In terms of energy efficiency, results compare quite favourably with other practices. The indicator of impact will ultimately be whether or not brickmakers invest in and adopt the technology.

Environment: more to be done

The environmental impact of such technological interventions, meanwhile, demands greater attention. Though no quantitative environmental impact assessments (EIAs) were carried out in the course of the project, it is possible to give a subjective account of the effect of the technologies developed. The use of waste material – sawdust and used engine oil – as fuel raises some crucial questions. Sawdust is more straightforward to analyse. Where it is used as a fuel, trees that would otherwise be felled to provide wood are conserved. Moreover, smoke output is generally reduced, because less wood is used when sawdust is incorporated into the body of clay bricks as a fuel than when wood is used directly to 'bottom-fire' a kiln. Therefore, firing with sawdust is more efficient. Also, it is probable that sawdust would otherwise be disposed of by unproductive burning or by dumping.

Obviously, the use of oil via a burner also saves trees that would otherwise be felled for fuel. The relative atmospheric pollution associated with burning oil as opposed to wood has not, however, been quantified in this project. If the technology is to be promoted further, it will be necessary to establish some form of control to avoid using oils contaminated with PCB (polychlorinated biphenyl) and chlorine, as these produce highly toxic emissions. Furthermore, the effect of burning this oil as opposed to its direct disposal on to the land or into rivers would need to be investigated. There is also a need to compare the trade-off between the extra or more damaging smoke caused by burning oil, coal or sawdust and the capture of emissions facilitated by the continued growth of trees that survive due to the adoption of these alternatives to fuelwood.

Information dissemination: getting the message across

Dissemination of the project's findings is continuing on a wide scale. This guidebook and a video on aspects of the work will, hopefully, further that aim. The published technical briefs, which have been referred to are already useful sources of information. Generally, the content is more suitable for engineers, scientists and field technicians than for direct access by the majority of small-scale brickmakers.

Distribution of information outputs has already been instigated via BASIN (Building Advisory Services Information Network). This has resulted in follow-up calls and requests for additional copies, notably from CRATerre, and organizations in India, Canada and Russia. ITDG's Technical Enquiries Unit has also responded to a number of enquiries. Articles in *Science, Technology and Development*, the DFID newsletter and on their website have also generated requests for information. The BBC World Service broadcast a feature on ITDG's work on brickmaking in Zimbabwe. Increased awareness of the key issues raised during the project has been generated through the media, especially in Peru, where the press has featured the project on several occasions.

There are two instances that are additional indicators of success. First, the Ministry of Energy of the Government of Peru uses ITDG as its source of information to compile statistics about the sector. Second, brickmakers in Piura are able to access funds from the Banco Materiales, a parastatal funding organization, because, due to their alliance with ITDG, they are using alternative fuels. The Banco Materiales is keen to assist activities that arrest deforestation.

10
GUIDELINES FOR PARTICIPATORY PROJECTS

FINALLY, IT IS TIME to consolidate guidelines for participatory technology transfer and development projects, taking into account all the factors that have been mentioned. Thus far, the text has ranged over a number of issues and attempted to present a wide variety of observations and insights. Focusing on the specific project with brickmakers as a case study has allowed some of the considerations involved with running a project to emerge naturally. Meanwhile, drawing on experiences outside this specific project has, hopefully, helped to broaden the perspective.

A number of the points raised and issues addressed do not need to be revisited in detail or rejustified in this concluding chapter. Nevertheless, it is worth reiterating that the search for principles to guide the process of participatory technology transfer and development, which has continued throughout the book, has yielded much that is of value. The elements of successful technology transfer, identified in a literature search and confirmed by ITDG's experience in Zimbabwe, were further confirmed by project experience in Peru and Ecuador. Moreover, the Ten Commandments formulated by IT Peru were found to hold good for the duration of the project. Add to these elements the observations of Jorge Marquina and Alfredo Barriga, and the reader will already have a reasonable basis for identifying valid guidelines.

So, now it is time to get to the crux of the matter, extending the principles mentioned in order to propose operational guidelines for participatory technology transfer and development projects and programmes. The questions are really: How do you plan, manage and enact a participatory project; what are the difficulties and the potential pitfalls; and how can they be addressed? With these questions in mind, the following discussion concentrates more on the methods of participatory working in technology transfer projects than on the technologies themselves: these have been discussed in the previous chapter.

Formulation: the project proposal and work plan

Before launching into definitive guidelines, it is worth presenting some ideas for project formulators, managers, instigators and funders to consider. This discussion focuses on the vital preliminary stage of formulating an intervention – a stage that is often neglected or 'short-changed', and which can make or break an intervention.

Formulating a project proposal, work plan and budget will be the point where most interventions start in earnest. Experience suggests that this is also a stage where many mistakes can be made. Planners must be acutely aware that they will be held to the proposal:

the activities in the work plan will have to take place, the books will have to be balanced. An ill-conceived, over-ambitious or poorly constructed proposal can doom a project from the outset.

Before a proposal can be written, a 'suitable' intervention must be identified and its validity researched to a significant extent. Ideally, a project proposal will be based on the expressed needs of the clients or beneficiaries. Furthermore, that need and the potential to address it should be assessed. The suitability of the proposing organization to intervene and instigate a project will also need to be evaluated: do we have the right experience, knowledge and skills? This self-evaluation should include an identification of the allies that will be necessary to ensure the intervention's success, including preliminary discussions with those allies. Pragmatically, the decision to formulate a project proposal must also be influenced by an assessment of the funds that are available. Are there donors with an interest in the sector where work is to be proposed?

Finding the right donor

Those formulating participatory projects have their work cut out. In many cases, the bureaucratic obstacles seem insurmountable. NGOs engaging in fieldwork are in a peculiar position; they seem to be serving two masters: the clients whose expressed needs are being inquired into, and the donors who have their own funding criteria.

NGOs obtain funding from a variety of sources, and each donor will have its own agenda and require funds to be applied for in terms and in forms dictated by this agenda. The task for the applicant is to find the right donor: one with an agenda that most nearly matches that of the NGO and the group or community it is seeking to assist and represent. The process of obtaining funding, as all NGOs know, can be lengthy, expensive in itself and exhausting. Approaching an unsympathetic donor with a proposal for a participatory project will be a terrible waste of time and energy. On the other hand, unless donors are made aware of the merits of participation and long-term commitment they cannot be expected to modify restrictive funding procedures.

Naturally, donors want to know how their money is to be spent and what the project will achieve. Unfortunately for participatory projects, this often means completing lengthy and complex application forms that are likely to demand the setting of targets, the achievement of measurable goals and the production of set outputs. The rigid strictures this can place on a project that is designed to develop dynamically with the community can serve to undermine the principles of participation. Private sector or corporate donors, particularly, but not exclusively, may be more interested in short-term, visible results: they want to see the return on their investment. An empowered community in a position to define their technology needs may not make for such obviously good publicity as a photograph of a smiling group of people standing around an imported – and wholly inappropriate – brick extruding machine.

There is often a clear difference between the priorities of small-scale producers in the developing world and those seeking to work with them, both project instigators and donors. From the experience of the brickmaking project in Latin America, for example, energy efficiency and environmental issues would not, in the first instance, be the principal concern of small-scale brickmakers – they are more concerned about reducing costs (see Box 8.7). Such concerns were, nevertheless, the motivation for the involvement of the project instigators and funders. It takes time to achieve consensus, to foster mutual understanding and establish shared definitions and goals. Not to dwell on this topic too long, finding a donor who is sympathetic to the aims of participation will obviously make life easier for project formulators and managers.

Accounting to the clients

The other 'master' is evidently the intended beneficiaries of the project. These beneficiaries – the NGO's clients – are not their paymasters. If this situation is not unique, it is at least unusual. Nevertheless, an NGO genuinely pursuing participation will want to make itself accountable to the group it has set out to assist. In the project with brickmakers in Latin America, project staff evolved an interesting way of thinking. They believed that the assistance – the service – they provided to brickmakers had to be so good that if the brickmakers had managed to get control of the project funds, they would choose to spend their money on it. Project staff believed this was a reasonable measure of their professionalism and worth. It is, at least, a perspective well worth considering.

Institutional allies

The other stakeholders to consider are the institutional allies that a project is likely to engage with – be they universities, local authorities or other NGOs. These organizations too will have an agenda: they will have expectations of the project that need to be taken into account.

Achieving a workable proposal

Finally, the time-limited nature of project funding – projects lasting perhaps two, three or four years – makes it less than ideal for participatory programmes of work. Combined with the often-inflexible performance demands of project proposal formats, this mitigates against participatory projects that involve long-term commitment, are dynamic in their approach and flexible in their ambitions. In brief, the imposed structure of some funding proposals is incompatible with participatory projects.

With this point in mind, the formulation of a project proposal in the ideal world would involve instigators, clients, donors and allies all meeting together. In the real world, however, the instigating NGO is much more likely to have to meet some of the other stakeholders in isolation. The task is then to represent faithfully the views of, for example, the clients to the donors. Indeed, it is unlikely in most instances that donors will ever meet either the clients or project allies face to face.

In writing a proposal, though, the instigator should involve – as far as possible – the clients and the project allies. As mentioned already, it is also a prerequisite to have as much knowledge as possible of the sector in which the intervention is planned: the socio-economic, politico-cultural and environmental context in which the work will take place. To that end, market research, social surveys and potential environmental impact assessments may all be necessary. The budget and work plan will have to include time and money for staff training and research and development, including a contingency for the cost of trials that may fail. In the final analysis, the project proposal has to be extremely well grounded, thoroughly thought out and realistic.

Summary: consolidating guidelines

Having given special attention to the often-neglected aspect of formulating a proposal, it is time to move on to consolidating the information that has been presented in this book and presenting a 'check list' of guidelines, which project instigators can refer to. Whether or not it is explicitly stated in the following points, the participation of stakeholders and allies is assumed in all stages of an intervention, from formulation through implementation to continuation or termination. That point notwithstanding, the degree of participation will, as has already been stated, depend on the specific situation.

The following guidelines are proposed sequentially – as far as that is possible. In reality, things never break down as neatly, and so adjustment of the order of steps and stages will inevitably be necessary. Furthermore, the list obviously will not be comprehensive for every intervention. Neither is it intended to be prescriptive – all projects differ and some of the proposed guidelines may not be applicable in every case. Nevertheless, it will be useful for instigators to consider the list and examine how well a particular project fits in with the guidelines. Allowing enough time for each stage to take action in accordance with the guidelines is essential, but it inevitably involves the project manager in a difficult juggling act. There will be unforeseen setbacks (for example, the delays to trial firings caused by El Niño flooding in Peru and Ecuador) and things will change. Nevertheless, time has to be made available to consolidate research findings and training, repeat trials and communicate. The guidelines are presented as a list of steps:

- Evaluate the existing indigenous technology, assess the exogenous technology options, and decide upon the scope for transfer and development.
- Study the potential impact of intervention on livelihoods, markets, the environment and any social group liable to be affected.
- Define priorities with stakeholders and allies: discuss expectations; establish 'common cause' and a vision; agree an agenda.
- Plan the intervention with the participation of all stakeholders and allies: decide upon roles and responsibilities; allow for on-going learning and dynamic development; and set realistic goals.
- Consider a pilot phase to test the project's approach and methodologies.
- Set up monitoring and evaluation procedures, including a system for clear and comprehensive documentation of the history of the intervention.
- Establish an internal and external communications strategy to include – as appropriate – networking, dissemination and advocacy. Make outputs accessible to the intended audience, paying attention to employing appropriate language and media.
- Examine how the intervention can be made sustainable, and develop a project continuation or exit strategy with stakeholders.
- Make the enhancement of knowledge and skills the key element of the intervention.
- Identify the training and capacity-building needs of all stakeholders and allies, including staff within the instigating organization.
- Select training methodologies and mechanisms that are likely to be most effective and appropriate to needs.
- Review the background educational needs of stakeholders and decide what actions are required for them to gain maximum benefit from vocational training.
- Assess the support programmes that may be needed to augment technical training. For example: enterprise development, bookkeeping, business management, co-operative working practice, business legislation, marketing, consumer education, health and safety, etc.

- Conduct research and development locally within a sheltered environment; ensure technology is proven and that the project team is confident in it and competent in its application prior to dissemination.
- Research the need for institution building or institutional development within the sector where the intervention is taking place.

Further reading

Participation, training and technology development

Appleton, H. (ed.) (1995) *Do it herself: women and technical innovation*, ITDG Publishing.

Appleton, H., M.E. Fernandez, C.L.M. Hill and C. Quiroz (1995) 'Claiming and using indigenous knowledge', in *Missing links: gender equity in science and technology for development*, Gender Working Group, United Nations Commission on Science and Technology for Development.

Biggs, S. and G. Smith (1998) 'Beyond methodologies: coalition-building for participatory technology development', *World Development*, Vol. 26(2), pp. 239–248.

Blackburn, J. and J. Holland (eds) (1998) *Who changes? Institutionalizing participation in development*, ITDG Publishing.

'Bridging gaps through co-operation' (1997) *BASIN News*, No. 14, August.

Buatsi, S. (1988) *Technology transfer: nine case studies*, ITDG Publishing.

Chambers, R. (1997) *Whose reality counts? Putting the first last*, ITDG Publishing, London.

Davies, A. (1997) *Managing for a change: how to run community development projects*, ITDG Publishing in association with Voluntary Service Overseas.

Eade, D. (1998) *Capacity-building: an approach to people-centred development*, Oxfam Publications, Oxford.

Everts, S. (1998) *Gender and technology: empowering women, engendering development*, Zed Books, London.

Guidelines for transferring effective practices: a manual for south–south co-operation (1998) CityNet/UNDP/UNCHS.

Guijt, I. and L. van Veldhuizen (1998) *What tools? Which steps? Comparing PRA and PTD*, IIED.

Guijt, I. and M.K. Shah (1998) *The myth of community: gender issues in participatory development*, ITDG Publishing.

Harper, M. (1996) *Empowerment through enterprise: a training manual for non-government organizations*, ITDG Publishing.

Haverkort, B. and W. Hiemstra (eds) (1998) *Food for thought: ancient visions and new experiments of rural people*, Books for Change and Zed Books, London.

Hennen, L. (1999) 'Participatory technology assessment: a response to technical modernity?' *Science and Public Policy*, Vol. 26, No. 5, Beech Tree Publishing.

IDRC (1995) *Missing links: gender equity in science and technology for development*, International Development Research Centre, ITDG Publishing/UNIFEM.

Jeans, A., E. Hyman and M. O'Donnell (1990) *Technology: the key to increasing the productivity of microenterprises*, Development Alternatives for USAID, Gemini Working Paper No. 8.

Leurs, R.A. (1993) *Resource manual for trainers and practitioners of Participatory Rural Appraisal (PRA)*, University of Birmingham.

MacDonald, I. and D. Hearle (1990) *Communication skills for rural development*, Evans Brothers.

Margoulis, A., et al. (1992) *Selection of group exercises and games for PRA trainers*, IIED.

Marshall, K. (1983) *Package deals: a study of technology development and transfer*, ITDG Publishing.

McCabe, A., V. Lowndes and C. Sklecher (1997) *Partnerships and networks: an evaluation and development manual*, YPS.

Narayan D. and L. Srinivasan (1994) *Participatory development tool kit: training materials for agencies and communities*, World Bank.

New approaches to science and technology, co-operation and capacity building (1999) United Nations Publications.

Oyelaran-Oyeyinka, B. (1997) 'Technological learning in African industry: a study of engineering firms in Nigeria', *Science and Public Policy*, Vol. 24, (5), pp. 309–318.

Parker, A.R. (1993) *Another point of view: a manual on gender analysis training for grassroots workers*, UNIFEM.

'Participation, literacy and empowerment' (1998) International Institute for Environment and Development (IIED), PLA Notes No. 32.

'Performance and participation' (1997) International Institute for Environment and Development (IIED), PLA Notes No. 29.

Platt, L. and G. Wilson (1999) 'Technology development and the poor/marginalised: context, intervention and participation', *Technovation*, No. 19, Pergamon.

Powell, J. (1995) *The Intermediate Technology Transfer Unit: a handbook on operations*, ITDG Publishing.

Pretty, J., I. Guijt, J. Thompson and I. Scoones (1995) *Participatory learning and action: a trainer's guide*, IIED.

Same platform, different train: the politics of participation (1998) The Corner House Briefings, No. 4.

Schubert, B., A. Addai and S. Kachelriess (1994) *Facilitating the introduction of a participatory and integrated development approach (PIDA) in Kilifi District, Kenya: Volume 2: from concept to action: a manual for trainers and users of PIDA*, Humboldt University.

Schwarz, R.M. (1994) *The skilled facilitator: practical wisdom for developing effective groups*, Jossey-Bass.

Srinivasan, L. (1990) *Tools for community participation: a manual for training trainers in participatory techniques*, UNDP.

Sweetman, C. (ed.) (1998) *Gender, education and training*, Oxfam, Oxford.

Technology and development: strategies for the integration of gender (1997) Tool/Tool Consult.

van der Bliek, J. and L. van Veldhuizen (1993) *Developing tools together*, ETC Foundation.

van Veldhuizen, L., A. Waters-Bayer and H. de Zeeuw (1997) *Developing technology with farmers: a trainer's guide for participatory learning*, Zed Books, London.

Varghese S. and J. Seed (1991) *Training of gender trainers workshop*, Gender and Development Unit, Oxfam.

Wijethilake, S., P. Fernando and H. Appleton (in press) *Discovering technologists: women's and men's work at village level*, IT, Sri Lanka.

World Employment Report (1998–99) Employability in the global economy – how training matters (press summary), International Labour Organisation.

Brickmaking, building materials, shelter and energy

Beamish, A. and W. Donovan (1989) *Village-level brickmaking*, Vieweg & Sohn (Aus der Arbeit von GATE), Braunschweig/Wiesbaden.

Der-Petrossian, B. (1997) *Global overview of construction technology trends: energy efficiency in construction*, UNCHS (Habitat), Nairobi.

Dudley, E. and A. Haaland, (1994) *Communicating building for safety: guidelines for methods of communicating technical information to local builders and householders*, ITDG Publishing, London.

Hammond, A.A. (1997) *Small and medium-scale brick and tile production in Ghana: overview, technology alternatives and energy alternatives*, GATE Wall Building Case Studies, GATE/GTZ-Eschborn.

International Labour Organisation (1984) *Small-scale brickmaking*, ILO Technical Memorandum No. 6.

Jones, T. (1995) *Brick clamps*, GATE Wall Building Case Studies, GATE/GTZ-Eschborn.

Jones, T. (1996) *The basics of brick kiln technology*, Vieweg & Sohn (Aus der Arbeit von GATE), Braunschweig/Wiesbaden.

Mason, K. (1997) 'Energy efficiency in small-scale brickmaking: experience of the Intermediate Technology Development Group (ITDG) in Zimbabwe', *Science Technology and Development*, Vol. 15, No. 1, pp. 162–173, Frank Cass, London.

Mason, K. (2000) *Assessing the technical problems of brick production: a guide for brickmakers and fieldworkers*, ITDG UK/Peru (available in English and Spanish).

Mason, K. (2000) *How to measure the energy used to fire clay bricks: a guide for brickmakers, fieldworkers and researchers*, ITDG UK/Peru (available in English and Spanish).

Mason, K. (2000) *Ten rules for energy efficient cost effective brick firing: a guide for brickmakers and fieldworkers*, ITDG UK/Peru (available in English and Spanish).

Merschmeyer, G. (1989) *Basic know-how for the making of burnt bricks and tiles*, Miserior.

Miles D. and R. Neale (1991) *Building for tomorrow: international experience in construction industry development*, International Labour Office (ILO).

Peltenberg, M., F. Davidson, H. Teerlink and P. Wakely (1996) *Building capacity for better cities: cases*, Institute for Housing Studies Development Planning Unit.

Ruskulis, O. (1999) *Selected bibliography on brickmaking in developing countries*, GATE Wall Building Bibliography, GATE/GTZ–Eschborn.

Schilderman, T. (2000) *Sustainable small-scale brick production: a question of energy? International research experience in brick production*, ITDG UK/Peru (available in English and Spanish).

Spence, R., et al. (1991) *Energy for building*, UNCHS (Habitat), Nairobi.

Spence, R.J.S. and D.J. Cook (1983) *Building materials in developing countries*, Wiley, Chichester.

Stulz, R. and K. Mukerji (1993) *Appropriate building materials: a catalogue of potential solutions*, 3rd edn, SKAT Publications and ITDG Publishing (also available in French and Spanish).

UNHCS (1984) *Sites and services schemes: the scope for community participation*, Training module, UNCHS (Habitat).

UNHCS (1986) *The role of women in the execution of low-income housing projects*, Training module, UNCHS (Habitat).

UNHCS (1991) *Guide for designing effective human settlements training programmes*, UNCHS Training materials series.

Websites

www.gtz.de/basin/gate *BASIN's site has a range of useful technical briefs on brick-firing and case studies on transferring brick-firing technologies, as well as much else on building materials and construction methodologies.*

Information on ITDG's programmes in the following sectors: manufacturing, food production, agro-processing, energy, building materials and shelter, disaster prevention and mitigation, transport and mining.

ITDG Publishing produces books on brickmaking, participatory technology development and other aspects of development. It also has an online mail order service of around 3000 development book titles.

References

Appleton, H. (1994) 'Ownership through participation', *Appropriate Technology*, Vol. 21, No. 1, June.

Aristide, J.-B. (2000) *Eyes of the heart: seeking a path for the poor in the age of globalization*, Common Courage Press.

Bairiak, J. (1999) *Utilization of bagasse in brickmaking: R&D in Sudan*, Wall Building Technical Brief, GATE.

Barriga, A., et al. (1992) *Brick and lime kilns in Ecuador: an example of woodfuel use in Third World small-scale industry*, Stockholm Environment Institute.

Biggs, S.D. (1995), 'Participatory Technology Development: reflections on current advocacy and past technology development', paper for workshop on PTD at the Institute of Education, Bedford Way, London, 23 March.

Browne A.W. (1981) 'Appropriate technology and the dynamics of village industry', *Institute of British Geographers*, New Series, No. 6.

Browne, A.W. (1982) 'Rural development in Botswana: the role of small-scale industries', *Geography*, Vol. 67.

Chigaru, P. (1999) *Greenline*, Issue No. 17, Zimbabwe.

Floorman, S.C. (1994) *The existential pleasures of engineering*, second edition, St Martin's Press, New York.

Guijt, I. and L. van Veldhuizen (1998) *Which tools? Which steps? Comparing PRA and PTD*, IIED.

Hammond, A.A. (1997), *Small and medium scale brick and tile production in Ghana: an overview*, GATE Wall Building Case Studies, GATE/GTZ-Eschborn.

Haverkort, B., J. van der Kemp and A. Water-Bayer (1991) *Joining farmer's experiments: experiences in PTD*, ITDG Publishing, London.

Hope, A. and S. Timmel (1999), *Training for transformation: a handbook for community workers*, Books 1, 2 and 3, Mambo Press, 1995, Book 4, ITDG Publishing, London.

Jazdowska N. (1997) *Review of technology transfer and small-scale brickmakers*, ITDG Zimbabwe, Harare.

Jones, T. (1997) *The vertical shaft brick kiln: a problematic introduction into Pakistan*, Wall Building Case Study, BASIN.

Khandker, S.R. (1998) *Fighting poverty with microcredit*, Oxford University Press.

Koopmans, A. and S. Best (1993) *Status and development issues of the brick industry in Asia*, Regional Wood Energy Development Programme in Asia of the FAO of the UN, Bangkok, (GCP/RAS/131/NET Field Document No. 35).

Lim, C.P. (1978) *Choice of manufacturing technology in the leather shoe and brick industries in Malaysia*, ILO, Working Paper 42.

Lowe, L. (1997) *Earthquake resistant housing in Peru*, ITDG Publishing.

Lowe, L. (2000) 'Project completion report', *DFID PCR*, No. 6483.

Lowe, L. (2001) *Building in partnership: participatory technology development*, ITDG, Rugby.

Mandela, N. (1999) 'The Sacred Warrior', *Time*, December 31.

Manser, W.A.P. and S. Webly (1979) *Technology transfer to developing countries*, The Royal Institute of International Affairs,

Mason, K. (1993) 'Coal-fired, small-scale brickmaking in Zimbabwe', *BASIN News*, Issue No. 6, July.

Mason, K. (1994) *Evaluation of clay brick burning in Cajamarca and Alto Mayo region*, ITDG Peru, April.

Mason, K. (1997) 'Brickmaking in Zimbabwe', *Science, Technology and Development*, Vol. 15, No. 1, April.

Mason, K. (2000) *Ten rules for energy-efficient cost-effective brick firing: a guide for brickmakers and fieldworkers*, ITDG UK/Peru (available in English and Spanish).

Murwira, K., H. Wedgwood, C. Watson, E.J. Win and C. Tawney (2000) *Beating hunger – the Chivi experience: a community-based approach to food security in Zimbabwe*, ITDG Publications, London.

Mutsambiwa, F. (1993) 'Is commercial viability in small-scale brickmaking attainable or a far cry?', *BASIN News*, Issue No. 6, July.

Newsweek (2000) quoting from *Encyclopaedia of the Future, Britannica Book of the Year, two centuries of population growth, 1950–2150, International Historical Statistics 1750–1980, AD 2000* Global Monitor.

Pretty, J. (1974) *Environmental assessment and participation: some challenges and dangers*, IIED.

Rutherford, S. (1995) *The biography of an NGO: empowerment and credit in rural Bangladesh*, Association for Social Advancement.

Santikarn, M. (1981) *Technology transfer*, Singapore University Press.

Schilderman, T. (1998) 'Sustainable materials production: a question of energy', a paper given at the CECAT conference, Ecomateriales y Habitat Sostenible, in Havana, Cuba, November.

Schilderman, T., S. Modi and K. Tideman (1999) 'Evaluation report on the SDC action research project on Vertical Shaft Brick Kiln Technology in India', ITC unpublished report.

Schumacher, E.F. (1974) *Small is beautiful: a study of economics as though people mattered*, Abacus Edition, Sphere Books.

Smillie, I. (1986) *No condition permanent: pump priming Ghana's industrial revolution*, ITDG Publishing.

Stewart, C.T. and Y. Nihei (1987) *Technology transfer and human factors*, Lexington Books.

Stewart, F. (1978) *Technology and underdevelopment*, Macmillan.

Tawodzera, P. (1994) 'Kurehwaseka Brickmaking Co-operative: a marginal small-scale brick producer on the road to success and recognition', *BASIN News*, Issue No. 8, July.

UN (1986) Advisory Service on Transfer of Technology.

UNCHS (1989) *Community credit mechanisms: training module*, UNHCS (Habitat).

UNCHS (1996) *An urbanising world – global report on human settlements*, Oxford University Press.

Van Ginneken, W. and C. Baron (eds) (1984) *Appropriate products, employment and technology: case studies on consumer choice and basic needs in developing countries*, Macmillan.

Versluyen, E. (1999) *Defying the odds: banking for the poor*, Kumerian Press.

Wilkinson, R. (1999) *Energy constraints in peri-urban areas around Kumasi and Hubli Dharwad*, Intermediate Technology Consultants.

Wolf, H. (1994) *Appropriate Technology*, Vol. 21, No. 1, June.

Appendix 1
Project time frame

1996

Technological needs assessment and agreeing objectives with brickmakers in Peru

Work begins in the areas La Huaca, Piura; Cajamarca; and Ayacucho. This involved visiting each area to carry out surveys for baseline socio-economic and technical information, and establishing working agreements with the producers.

Developing a methodology for measuring specific energy use of kiln

This new methodology (see Chapter 7) was tested out on six firings in Zimbabwe.

Literature search to identify possible energy-efficient methods for Peru

Making contact with relevant institutions

These included: ESPOL, Ecuador; PROLENA, Nicaragua; ULOCEPI Project, Bolivia; and DCR/FEAGRI/UNICAMP, Brazil.

1997

International workshop, February 1997

Seven case studies were presented relating to fuel usage in brickmaking in several countries of Latin America and Zimbabwe. A new project partner emerges: Dr Alfredo Barriga of ESPOL, Ecuador.

Experimentation begins at La Huaca, Peru

A coal-and-firewood hybrid kiln is fired in March.
Workshop conducted at La Huaca at which the Ten Rules for energy-efficient brickmaking were derived (March).
Experimentation continues in La Huaca during 1997 with coal and coal-dust briquettes. Firings are temporarily abandoned during the heavy rains caused by El Niño in November.

Ayacucho, Peru, September–December

Visit by brickmakers from La Huaca, followed by trials of new fuels. Trial kilns include: complete substitution of fuelwood by coal, incorporating coal-dust briquettes; semi-bituminous coal and wood mixes; anthracite supplemented by briquettes.
Different-sized kilns are experimented with: kilns with capacities of 9000, 22 000, 24 000 and 31 000 bricks are compared.
For the first time, the energy efficiency of all of these is measured using the methodology developed by the project. As well as the energy used, the internal temperatures are measured using 'Bullers bars'. The calorific value of fuels is tested at a local laboratory.

Establishing links with training organizations in Peru

Discussions have revealed that some organizations are more interested in working with large-scale industry; SENCICO (works with artisans in the building industry) is still interested in working with small-scale brickmakers, however, and in continuing ITDG's efforts in Peru after the end of the project.

Formalizing agreement with ESPOL, Ecuador

ITDG Peru's agreement with ESPOL is finalized, and surveys are undertaken to assess the available energy resources and to monitor the energy efficiency of existing practices. Trial kilns planned in Guayaquil, Ecuador; but these are postponed due to poor weather (El Niño).

1998

International workshop, Lima, Peru, February

Participants include brickmakers; coal suppliers; and representatives from ITDG Peru, Zimbabwe and UK, and from technology transfer organizations. The workshop considered the equipment required and the methodology for technology transfer to be applied during the project.

Ayacucho, Peru

Experimental firings continue using the project's demonstration kiln. Three firings employ coal-dust briquettes, with the minimum of fuelwood. Ayacucho brickmakers visit Trujillo to see oil-burner demonstrated.
Firings in Ayacucho and La Huaca are filmed, in order to produce a training video.
By the end of 1998, activities on the project kiln in Ayacucho proceed without further fuel or labour inputs from the project.

La Huaca, Peru

Following the period of flooding, brickmakers resume activities.
Brickmakers form an association in March – the Yawar Huaca Brickmakers Association.
Ten members of the Association visit Ayacucho in May to see the use of briquettes demonstrated.
A Scotch kiln of 13 000 bricks is constructed in La Huaca and fired with briquettes.
Banco de Materiales assists brickmakers in the Association with loans for repairing kilns following rains (July).

Following brickmakers' visit to Trujillo to see oil-burner demonstrated, and return visit by Trujillo brickmaker, in November, brickmakers purchase an oil-burner to ignite their all-briquette kiln, thus eliminating the need for fuelwood altogether.

Cajamarca

Work recommences with brickmakers substituting coal for fuelwood.

Developing equipment

Two manually operated machines, the Pendulum Clay Crusher and Bulley Clay Extruder, are purchased in the UK and shipped to Peru.
These are tested in May: the crusher works on clays from La Huaca, but not from Ayacucho, which contain hard impurities. The extruder's capacity is considered to be too meagre, but serves as a reference for developing an extruder driven by a compression screw and one driven by a motor.

Links with training organizations

Staff members from SENCICO (Peru) and ITACAB (operating in Bolivia, Colombia, Ecuador, Spain, Panama, Venezuela and Peru) attend project activities to aid technology transfer across the region.

1999

La Huaca, Peru

Three brickmakers of the Yawar Huaca Association travel to Trujillo in March to make contact with coal suppliers; from now on coal deliveries can be arranged by phone.
A coal store is established on a brickmaker's land.
Firings in May involve complete substitution of fuelwood by coal, coal briquettes and oil-burner. As firings continue throughout the year confidence is gained in the use of the oil-burner and project assistance is no longer required.
Six trial firings also carried out to test the indigenous practice of mixing rice husk into bricks in various proportions (April–June).
Brickmakers also experimented with mixing sawdust into bricks.
Work to formalize the brickmakers' association begun in order to manage the kilns and equipment left by the project as a business concern once the project activities in the area are finished.

Cajamarca, Peru

Brickmakers begin to substitute coal for fuelwood, following training by brickmakers of La Huaca.
Trial firing of kilns using coal, coal-dust briquettes and oil-burner takes place successfully in November, still with external assistance.

Ayacucho, Peru

Last firing on the project kiln (February) is not so successful, and leads brickmakers to seek assistance from Yawar Huaca Association in using the oil-fired burner. By the end of 1999 brickmakers have completed the production of their own oil-burner, and have tried out firing a kiln with coal, coal briquettes and the oil-burner successfully

Chimborazo, Ecuador

Two firings take place, incorporating sawdust into the mass of the raw bricks (March). The strength of bricks is tested.

Riobamba, Ecuador

The oil-burner technology that has proved successful in Peru is introduced to Riobamba in November. Results achieved while working without external assistance are not yet satisfactory, however. Brickmakers also try incorporating sawdust into the bricks, with good results. Organizational aspects of the producers are addressed.

Equipment

A manual extruder and a manual clay mixer are constructed and tested.

Regional training workshop, Riobamba, Ecuador, November

For brickmakers and project staff to consolidate technology adoption.

2000

International workshop, February

For brickmakers, technicians and professionals to look back on project implementation and consider future opportunities.

Appendix 2
Energy monitoring forms

The following two monitoring forms demonstrate the specific firing energy that can be achieved under different conditions. The kiln monitored by Form 01 has coal dust substituted for some of the wood. The kiln monitored by Form 14 demonstrates the efficiencies that can be achieved with a larger kiln (15 000 bricks) and by total substitution of fuelwood by coal and oil.

The final, blank monitoring form may be photocopied by readers for their own use.

Energy consumption of brick firing processes 01		
NAME OF PRODUCER Victor Carmen	LOCATION /ADDRESS La Huaca, Paita, Piura, Perú	DATES OF FIRING Start 17 May 1997 15:45 Finish wood 18 May 1997 05:00 Finish coal 18 May 1997 17:00
TYPE OF CLAMP/KILN 2 tunnel, Scotch kiln Size: 3.25 m × 2.45 m × 3.3 m	TYPE OF FUEL Algarrobo wood and semi-bituminous coal dust	MASS OF FUEL USED (kg) 2270 (algarrobo) 1400 (coal)
CALORIFIC VALUE (kJ/kg) (i) Algarrobo Gross = 17 555 kJ/kg Net = 16 310 kJ/kg Moisture cont. = 10% (ii) Coal Gross = 17 061 kJ/kg Net = 15 547 kJ/kg Moisture cont. = 3.28%	NO. OF GREEN BRICKS 6358	MASS OF BRICKS wet = 4.20 kg dry = 4.11 kg fired = 3.75 kg
BRICK MOISTURE CONTENT 2.14%	METHOD OF FORMING Slop moulding	WEATHER CONDITIONS Dry, hot; gusting light wind
CALCULATION OF KILN EFFICIENCY Mass of green brick = 26 703.6 kg Total moisture content = 572.22 kg Drying energy = 1 482 622.02 kJ Wood energy = 37 023 700.00 kJ Coal energy = 21 765 800.00 kJ Gross energy = 58 789 500.00 kJ Firing energy = 57 306 877.98 kJ Mass of fired brick = 23 842.50 kg Specific firing energy = 2.40 MJ/kg		QUALIFYING INFORMATION (i) Vitrification temp. = 1150°C Category of soil = high temp. (ii) Bullers ring no. = 55 (iii) Max. temp. = 970°C (iv) Firing time = 38 h 15 m

COMMENTS

Initial firing was too rapid to take full advantage of the effect of placing coal in layers. 90% good bricks, 10% under-fired and broken. All dust coal was fired. Coal-dust firing time was estimated. Data reading taken on site by Mario Jara.

Signature, date, organization and contact address:
Emilio Mayorga/Teodoro Sánchez, 31 October 1997. ITDG-Perú, Lima.
Fax (511) 4466621, e-mail: teo@itdg.org.pe

Energy consumption of brick firing processes 14

NAME OF PRODUCER	LOCATION /ADDRESS	DATES OF FIRING	
ITDG Kiln	La Huaca, Paita, Piura, Perú	Start	15 April 1999 11:00
		Finish oil	15 April 1999 14:30
		Finish briquette	16 April 1999 18:00
		Finish coal	19 April 1999 18:00
TYPE OF CLAMP/KILN	TYPE OF FUEL	MASS OF FUEL USED (kg)	
2 tunnel, Scotch kiln	Anthracitic coal dust	3400 (coal)	
Size: 3.6 m × 3.0 m × 4.0 m	Oil	111.43 (oil) (139.30 l)	
CALORIFIC VALUE (kJ/kg) (i) Coal Gross = 28 099.00 kJ/kg Net = 26 533.68 kJ/kg Moisture cont. = 4.01% (ii) Oil Gross = unknown Net = 25 390.00 kJ/kg Moisture cont. = unknown	NO. OF GREEN BRICKS 15 000	MASS OF BRICKS wet = 4.20 kg dry = 4.11 kg fired = 3.78 kg	
BRICK MOISTURE CONTENT	METHOD OF FORMING	WEATHER CONDITIONS	
2.14%	Slop moulding	Hot and dry	
CALCULATION OF KILN EFFICIENCY Mass of dry brick = 61 650.00 kg Total moisture content = 1 350.00 kg Drying energy = 3 497 850.00 kJ Oil energy = 2 829 207.70 kJ Gross energy = 93 043 719.70 kJ Firing energy = 89 545 869.70 kJ Mass of fired brick = 56 700.00 kg Specific firing energy = 1.58 MJ/kg		QUALIFYING INFORMATION (i) Vitrification temp. = 1150°C Category of soil = high temp. (ii) Max. temp. = 970°C (iii) Firing time = 103 h	

COMMENTS

Brick quality: mostly well-fired. Coal-dust firing time was estimated. Data reading taken on site by Jorge Marquina.

Signature, date, organization and contact address:
Emilio Mayorga/Teodoro Sánchez, 31 October 1997. ITDG-Perú, Lima.
Fax (511) 4466621, e-mail: teo@itdg.org.pe

Energy consumption of brick firing processes		
NAME OF PRODUCER	LOCATION /ADDRESS	DATES OF FIRING Start Finish (each fuel)
TYPE OF CLAMP/KILN	TYPE OF FUEL	MASS OF FUEL USED (kg)
CALORIFIC VALUE OF EACH FUEL (kJ/kg) Gross: Net: Moisture cont.:	NO. OF GREEN BRICKS	MASS OF BRICKS wet: dry: fired:
BRICK MOISTURE CONTENT (%)	METHOD OF FORMING	WEATHER CONDITIONS
CALCULATION OF KILN EFFICIENCY Mass of wet brick (kg): Mass of dry brick (kg): Total moisture content (kg): Drying energy (kJ): Total energy (kJ): Firing energy (kJ): Mass of fired brick (kg): Specific firing energy (kJ):		QUALIFYING INFORMATION (i) Vitrification temp.: Category of soil: (ii) Bullers bar temp. No. and/or Avg. firing temp. (iii) Firing time (hours):

NOTES:

COMMENTS:

RECOMMENDATIONS:

SIGNATURE, DATE, ORGANIZATION AND CONTACT ADDRESS:

Appendix 3
A comparison of calorific values for different fuels

Table 1 Indicative calorific values for some common fuels

Fuel	Energy value (MJ/kg)
Commercial butane	58
Diesel fuel	44
Heavy fuel oil	42
Charcoal (2% moisture)	33
Anthracite coal	29
Waste engine oil	25
General purpose coal (non-coking)	23
Dry sawdust	18
Wood (15% moisture)	15

Table 2 Approximate conversion of energy units

	TPE	MJ	kWh	TCE
Relative energy value	1	41 868	11 630	1.43

Table 3 Approximate conversion of units for calorific values

	MJ/kg	cals/g	Btu/lb	kWh/tonne
Relative calorific value	1	239	430	278

Note: MJ = megaJoules, kg = kilogram, TPE = tonnes of petrol equivalent, kWh = kilowatt hour, TCE = tonnes of coal equivalent, cals/g = calories per gram, Btu/lb = British thermal units per pound.

www.ingramcontent.com/pod-product-compliance
Ingram Content Group UK Ltd.
Pitfield, Milton Keynes, MK11 3LW, UK
UKHW060455150426
5217IPUK00028B/2086